CW00500960

In the Looking Glass

In the Looking Glass

Mirrors and Identity in Early America

Rebecca K. Shrum

Johns Hopkins University Press
Baltimore

Published 2017
Printed in the United States of America on acid-free paper
9 8 7 6 5 4 3 2 1

Johns Hopkins University Press
2715 North Charles Street
Baltimore, Maryland 21218-4363
www.press.jhu.edu

Library of Congress Cataloging-in-Publication Data

Names: Shrum, Rebecca K. (Rebecca Kathleen), 1972– author.
Title: In the looking glass : mirrors and identity in early America / Rebecca K. Shrum.
Description: Baltimore : Johns Hopkins University Press, 2017. | Includes bibliographical references and index.
Identifiers: LCCN 2016046947| ISBN 9781421423128 (hardcover : acid-free paper) | ISBN 142142312X (hardcover : acid-free paper) | ISBN 9781421423135 (electronic) | ISBN 1421423138 (electronic)
Subjects: LCSH: United States—Social life and customs—To 1775. | United States—Social life and customs—1775–1783. | United States—Social life and customs—1783–1865. | United States—Social conditions—To 1865. | Mirrors—Social aspects—United States—History. | Identity (Psychology)—Social aspects—United States—History. | Race awareness—United States—History. | Sex role—United States—History. | Technology—Social aspects—United States—History. | Material culture—United States—History.
Classification: LCC E162 .S557 2017 | DDC 973.1—dc23
LC record available at https://lccn.loc.gov/2016046947

A catalog record for this book is available from the British Library.

Special discounts are available for bulk purchases of this book. For more information, please contact Special Sales at 410-516-6936 or specialsales @press.jhu.edu.

Johns Hopkins University Press uses environmentally friendly book materials, including recycled text paper that is composed of at least 30 percent post-consumer waste, whenever possible.

For Matthew Brady Shrum

Contents

In the Looking Glass

Introduction

Early one morning in late September of 1805, the young army officer Zebulon Pike awoke to discover that his expedition's flag, which had been hanging the night before on the deck of his ship, moored near what would become Fort Snelling, Minnesota, had disappeared. Irritated, he wondered "whether it had been stolen by the Indians, or had fallen overboard and floated away." Downriver from Pike's camp, Cetanwakanmani, the Mdewakanton Dakota chief with whom Pike had just negotiated to acquire nine square miles of land for the United States, found the flag in the water and, curious about whether it signified trouble for the Americans, headed north to investigate. After Cetanwakanmani arrived at the camp and saw that all was well, he told Pike about a series of events surrounding the flag's discovery. Another chief, Outarde Blanche, had come to Cetanwakanmani to exact revenge because his lip had been cut off in a previous altercation. The disfigurement anguished Outarde Blanche because "his face was his looking-glass" and it "was spoiled." The object Outarde Blanche identified as his "looking-glass" would be more familiar to us today as a mirror; it was a small piece of reflective glass made in Europe that he had likely acquired through trade with whites.[1] Cetanwakanmani and Outarde Blanche "were charging their guns and preparing for action when lo! the flag appeared like a messenger of peace sent to prevent their bloody purposes." Curious about the Americans' fate, the men stood down.[2]

Whether Outarde Blanche carefully planned the words he would use when confronting Cetanwakanmani or his utterance represented the first thing that came to mind in that moment of heated exchange, his declaration that "his face was his looking-glass" revealed the importance of the mirror for these two Native men: not only was it how Blanche had come to know his own appearance; it was a transformative object that he was able to imbue with feelings of revenge and fury that he directed at Cetanwakanmani. He expected his enemy to understand immediately the significance and importance of his invocation of the "looking-glass." Blanche's utterance relied on both men's intimate knowledge of what mirrors could do and how they worked to shape people's identities. The question that haunted Blanche was this: How would people perceive him now that his face—such a core component of his identity—had been compromised? From our vantage point more than two hundred years after this incident, we might ask: Why was it the looking glass that Blanche invoked to trigger the exchange between these two men? Cetanwakanmani could obviously see Blanche for himself. Why was it Blanche's own viewing of his visage that led to a potentially violent confrontation? Powerless to alter his appearance, Outarde Blanche invoked an object intimately linked to identity formation—the looking glass—as a way to give voice to what he had lost and to justify his revenge upon the man who had disfigured him.

Blanche had clearly developed deep knowledge about his appearance and a concomitant sense of self through using a mirror. He exclaimed that his face was his looking glass when expressing dismay about how the wound had altered not just his face, but because his face had come to represent his sense of who he was, his identity had been irrevocably changed as well. It has become a common linguistic trope—and a practice in everyday life—to use the face to signify the whole person. In doing so, as George Lakoff and Mark Johnson argue in *Metaphors We Live By*, "we perceive the person in terms of his face and act on those perceptions." Using the face to signify the whole person is more than a "referential device"; it also "serves the function of providing understanding." As an example, Lakoff and Johnson explain, "if you ask me to show you a picture of my son and I show you a picture of his face, you will be satisfied. . . . But if I show you a picture of his body without his face, you will consider it strange and will not be satisfied. You might even ask, 'But what does he look like?' "[3] This human tendency to associate other people's identities with their faces does not rely on the mirror, but the capacity to develop such an association for oneself does require a strong sense of what one's own face looks like. Mirrors made it possible for men and women across North America to develop over time a sense of self linked to their appearance, and it is that process to which this volume attends.

Blanche was not the only Native American who used the looking glass as a way to validate his personal experiences and sense of self. Osceola, a Florida Seminole leader who had been adamantly opposed to Indian removal, was tricked into capture in 1837 and imprisoned at Fort Moultrie, on Sullivan's

Island, South Carolina. During his captivity, several white painters, including George Catlin, traveled to Fort Moultrie to paint Osceola's portrait. Shortly after Catlin's departure, Osceola died. Later, Catlin received an account of his death, sent to him by Dr. Weedon, a surgeon who had witnessed it. According to Weedon, in his final moments Osceola prepared himself for death by calling for "his full dress." He then "put on his shirt, his leggings and moccasins—girded on his war-belt—his bullet-pouch and powder-horn, and laid his knife by the side of him on the floor." Finally, Osceola asked for his red paint and his looking glass. Someone held the glass before him so that he could paint "one half of his face, his neck and his throat." Osceola then painted his wrists, the backs of his hands, and knife handle, "a custom practiced when the irrevocable oath of war and destruction is taken."[4] As death approached, Osceola's final acts allowed him both to shape how he would look at the moment of his death and to have one final encounter with his mirror self.[5] This practice provided him a measure of dignity in preparing for an afterlife.

For people of African descent in nineteenth-century North America, mirrors were also inextricably linked to identity formation. A formerly enslaved woman, who went unnamed in the record of her 1930s interview, recounted the first time she saw herself in a mirror. Then a young enslaved girl, she had just been sold to a new family and was inside their home for the first time. She remembered how luxurious it was with "soft carpets" and "all the glasses around there." Approaching one of the mirrors for the first time, she remembered how she "just turned and looked and looked at myself, 'cause I had never seen myself in a glass before." In this memory of a young girl turning herself around in front of a mirror, we catch a glimpse of someone enthralled by her mirror self and eager to see as much of it as possible. Encountering her mirror self a second time that same day—after she had received new clothes and taken a bath—she stood before the looking glass, "pull[ing] up my dress and look[ing] at my pretty clean drawers and things."[6] Again, this elderly woman emphasized how, in these first encounters with her mirror self, she was eager to gain knowledge about her body and how she looked in those new clothes as she stood before the mirror. She relished the opportunity to engage with these images of herself and formed a mental picture of herself that remained with her into old age.

Jacob Green, another person of African descent, described a similar incident in the narrative of his life story. Born in 1813, Green recalled how, as a young man, he had once prepared to attend a dance for which his Aunt Dinah had starched the collars of his shirt, and he had "a good pair of trousers, and a jacket." After combing his hair, he "went and looked in an old piece of broken looking-glass" and thought to himself that "I was the best looking negro that I had ever seen." With twenty-four pennies in his pocket—amplified for effect by fifty large brass buttons—Green was ready to impress not only with his looks but with the jingling of his money. Green "thought what a dash I should cut among the pretty yellow and Sambo gals, and I felt quite confident, of course, that I should have my pick among the best looking ones, for my good clothes, and my abundance of money, and my own good looks—in fact, I thought no mean things of my self [*sic*]." Green's ability to know and assess his mirror self gave him confidence he would be appealing to those at the dance that night, and he carried that sense of self with him throughout his life.[7]

In these nineteenth-century accounts from people of Native and African descent in North America, we can clearly see men and women whose sense of themselves as individuals—their identity—relied in part on knowledge gained from encounters with the mirror self. Men and women of Native, African, and European descent in North America embraced the mirror as both a visual object and a tool capable of eliciting an emotive experience. Whites, however, also attempted to control the discourse that arose around mirrors. Whites claimed that they alone understood and used mirrors properly. Mirrors were tools of rational enlightenment, white men claimed, that could increase their mastery over themselves and their world. Moreover, whites believed that mirrors created evidence of African and Native inferiority as part of their larger project to differentiate themselves as "white" from those they deemed "red" or "black." People of European descent invented and then coalesced around this singular white identity as they interacted with other ethnic and racial groups through colonialism and empire building. By placing themselves at the pinnacle of America's developing racial hierarchy, whites of European descent in America were able to justify both slaveholding and territorial expansion through law, economic policies, and social norms. Importantly, though, mirrors were contested objects. People of African and Native descent resisted whites' categorization of them as inferior and embraced mirrors, in part, just as whites did, for the access they offered to vital information about the self. But encounters with the mirror self like those presented in my four opening vignettes are quite sparse, not surprisingly, given that the vast majority of what made it into the historical record from early American experiences was produced by and privileged the perspective of whites, who sought evidence of difference rather than similarity as part of their project of defining and shoring up the idea of "whiteness."[8] Yet African and Native voices broke through this agenda to provide direct evidence of how mirrors increased knowledge about the individual self. These disruptions form one of the core evidentiary bases of this volume.

In part, mirrors became contested objects in early America because their newness meant that their meaning was not yet fixed. Most of the mirrors into which early American men and women peered were a technological innovation of the period of the Western European Renaissance. In the early sixteenth century, Venetian craftsmen perfected a flat, clear glass and reflective coating that together produced accurately reflective glass mirrors. These new mirrors significantly surpassed the size and accuracy of earlier metal and glass mirrors, enabling men and women to encounter a clear image of their mirror self. Early Americans had uses for, and beliefs about, reflective surfaces that predated accurately reflective looking glasses and largely associated reflection with ritual and magic. The older beliefs and practices, as well as earlier metal and glass mirrors, existed in tension with these new mirrors, which early Americans embraced, in part, for the more accurate reflection of the self they provided. The accuracy of these mirrors was not, however, perfect. Because mirrors both render a three-dimensional object in a two-dimensional image and present to us a reflection of the self reversed from left to right (if you raise your right hand in front of the mirror, your mirror self appears to raise its left hand), observers always saw themselves differently in the mirror from how anyone else saw them.[9] While all mirrors, by definition, changed the mirror self in these ways, early American mirrors could also suffer from physical imperfections in both the flatness of the glass and the quality of the reflective coating that produced more noticeable dysmorphia. Despite these potential imperfections, the more accurate mirrors played a critical role in shaping a person's individual sense of self and came to be intimately linked to identity formation in all three cultures. Even as reflective glass mirrors created opportunities for new uses and meanings that linked reflection to identity formation, however, they also reinforced those preexisting systems of belief, related to ritual and magic, by making reflective devices more widely available.

Although many fields, including psychology, neuroscience, literature, and art history, have explored the significance of mirrors in the human story, the historical process by which mirrors shaped the identities of early Americans has remained largely unexamined.[10] But much work has been done that can help us understand the relationship of people to the objects of the material world. Mirrors belong to the category known as "material culture"—what the folklorist

Henry Glassie described as the "tangible yield of human conduct" and the historian Leora Auslander called "the class of all human-made objects." Much early work by historians that considered material culture was rooted in a consumerist framework, which focuses on what motivates people to become consumers and to purchase certain objects but does not linger on the relationships that can develop between people and their material goods before and after acquisition. Using Thorstein Veblen's late nineteenth-century theory of emulation, scholars in the consumerist framework first argued that elite white Americans acquired mirrors—and a wide range of other refined furnishings—in order to perform gentility. Thus, when ordinary white Americans acquired these same material goods, scholars understood that as an attempt to imitate their social betters. In recent years, however, a much more complex picture has emerged that acknowledges multiple motivations, including emulation, for the acquisition of goods as well as (un)expected meanings that develop over time, moving long past the point of purchase.[11]

To understand the significance of the goods we acquire and then live with over time, scholars began to consider objects in use in everyday life, what has come to be known as the "social life of things," and to examine how men and women imbued goods with particular meanings. T. H. Breen has explored how eighteenth-century consumers became producers of meaning for material goods, suggesting that when "a consumer acquired an object, he or she immediately produced an interpretation of that object, a story that gave it special significance." Amanda Vickery took a longer view to show how meanings ascribed to material goods developed over time as people interacted with them, investing their possessions with experiences and memories. Vickery also pointed out that we should not assume that material goods "carry the same social and personal meanings for all consumers." And Paul Clemens cautioned about the ability to uncover all of these meanings in any case, concluding in his study of the Middle Atlantic region in the mid-eighteenth to early nineteenth centuries that the wide range of "simple, useful, and practical" goods incorporated into ordinary households "were often more than merely useful or true necessities. They linked their owners to a transatlantic marketplace governed by European fashion and, to a degree scholars cannot fully measure, became a part of a personal identity fashioned from material possessions." What to an outsider might look like a set of unrelated goods may have reflected a series of conscious choices by a consumer, instilled with meaning understood fully only by that individual. The purchase and ownership of goods, then, is highly personal. We cannot capture the full range of meanings individual men and women ascribed to mirrors as part of their personal collections of material possessions, but close consideration of this one object will reveal patterns of meanings early Americans gave to mirrors. These patterns will enable us to better understand why mirrors became both vital and potentially subversive objects in early America.[12]

People do not only give meaning to material goods. The objects of the material world also make meaning for people and influence the human story. Material goods are conceptualized and created by men and women as they live their lives, interact with the material world, and imagine new items that have yet to be created. Objects enter a world that is already defined by this dynamic interaction of people and things and have the potential to refashion that world as they are deployed within and influence preexisting systems of meaning.[13] Thus it was that earlier reflective technologies—metal mirrors and poor-quality glass ones—inspired artisans to develop flatter, clearer glass and a more highly reflective coating. Once accurately reflective glass mirrors entered this dynamic environment, they shaped the individual identities of their owners and also influenced the early American experience in unanticipated ways.

The mirror has one unique feature that must be considered here: its capacity to reflect back a clear image of its user. All material objects have the capacity to be reflective figuratively, as a look around one's living room or a quick glance into a briefcase, backpack, or purse confirms. What we own tells us, and those around us, something about who we are. As historian and curator Carolyn Gilman describes it, "all artifacts are mirrors in their way."[14] Objects function as figurative mirrors by revealing facets of human identity and communicating messages of worth, desire, and belonging. But an actual mirror has both figurative and literal reflective capabilities. In the home of an eighteenth-century Boston merchant, a looking glass—like many of the other furnishings in that merchant's house—would have registered his wealth and social status by its size and finely crafted frame, as well as the prominence of its display in an important public room of his house. But the mirror would also have reflected back an image of him and his family, framed by the mirror's own opulence and its grandiose surroundings. Thus, mirrors could function like any other item bought and displayed in early America. Yet mirrors were the only object of the material world whose main purpose was to reflect clearly back to its user an image of the self. Looking into a mirror, owners or users could consider their physical appearance from moment to moment and draw conclusions about how others might see them. In this function the mirror was not just a part of the material world; it crafted the visual world on demand. As an item of visual culture, mirrors were objects through which people came to know themselves through sight.[15]

Early Americans gained this new knowledge about themselves as mirrors took up permanent residence on the walls of houses, in looking-glass cases, on dressing tables, in pockets, or in whatever place afforded a bit of privacy for valued personal possessions. Because looking glasses rendered men and women knowledgeable about what they looked like, something closely linked to identity formation, this object became intertwined with that identity.[16] How did this new visual knowledge affect early Americans' senses of themselves as individuals? What impact did it have on collective—even national—self-understanding?

Western thinkers long associated the development of an individual sense of self with the rise of modernity. The anthropologist Clifford Geertz described the modern individual as a "conception of the person as a bounded, unique, more or less integrated motivational and cognitive universe; a dynamic center of awareness, emotion, judgment, and action organized into a distinctive whole and set contrastively both against other such wholes and against a social and natural background."[17] Accurately reflective looking glasses—which provided, for the first time, realistic and regular visual access to the observer's own face and body—have long been presumed to have played a role in shaping the emerging, modern sense of individual selfhood Geertz described.[18] But these claims—about the rise of the modern individual and the mirror's role in that process—have been challenged in recent years. Scholars have countered that the idea of modern Western identity shaped around the sense of an individual self is a core part of a long-told fiction that, as historian Roy Porter argues, "embodies and bolsters core Western values." This fictional story juxtaposes "primitive" or "premodern" societies with "modern" ones. In "premodern" societies, identity formation was putatively fundamentally communal and collective because members were "so completely in the grip of supernatural and magical outlooks, ritual and custom as to preclude any genuine individuality." In "modern" ones, the identity that Western societies claimed for themselves, such an outlook had been overcome, at least for "literate, gifted, elite males," and "rationality could serve as the foundation-stone of the self-determining individual."[19] The period during which Western European societies supposedly became modern and birthed this individual sense of self has long been identified as the Renaissance, during which, as some have suggested, it was not a coincidence that accurately reflective glass mirrors also emerged.[20] But writing specifically about Renaissance mirrors, Debora Shuger has more recently argued that, despite the introduction of mirrors, the "Renaissance self lacks reflexivity, self-consciousness, and individuation, and hence differs fundamentally from what we usually think of as the modern self." Shuger challenges the idea that the introduction of accurately reflective glass mirrors had an immediate impact on Westerners' conception of themselves, concluding "that early modern selfhood was not experienced reflexively but, as it were, relationally."[21] In these ways the idea that Western societies birthed a modern individual sense of self during the Renaissance and that mirrors played a key role at that time in this process have both been brought into question.

In North America, whites attempted to tell this same fictional story, casting themselves in the role of the "rational," individual self while assigning the role of "primitive," communal, and collective selves to people of African and Native descent. Considering mirror ownership and use among people of African, Native, and European descent in North America enables us to see how people of European descent attempted to parcel out these roles as part of their larger project of constructing and solidifying "whiteness." What we will also see is that for people of European, Native, and African descent, the process by which the mirror shed its magical and ritual meanings to become a rational tool for increasing self-understanding was contingent and uneven. Evidence of magical practices remained in all three cultures throughout the nineteenth century. These magical beliefs and ritualistic uses for mirrors maintained at least a foothold (if not something more substantial) in all three cultures. At the same time, the ability of the mirror to shape an individual sense of self developed. Mirrors began to shape this sense of self for men and women from all three cultures as people first encountered looking glasses. Mirrors show how supposedly "modern" and "premodern" impulses could be joined in the same object and flourish there together for centuries. Early Americans increasingly understood themselves as individual men and women in part through regular encounters with their mirror selves, but they also retained earlier beliefs about and uses for reflective and deflective power. This study does not dispute the idea that there was "a fundamental shift in the West from 'tradition' to 'modernity' over the long term of the sixteenth, seventeenth, and eighteenth centuries," but, rather, it explores why, in North America, the European American project of developing and maintaining the idea of whiteness compelled whites to claim modernity for themselves and attempt to deny it to people of African and Native descent.[22]

I begin this study in Europe, with a short exploration of the history and development of mirror technology. We cross the Atlantic by the close of chapter 1 to understand how the mirror was initially viewed in the Euro-American context. Chapter 2 explores the rich array of archaeological, documentary, and linguistic evidence that allows us to observe how two seventeenth-century cultures—Native peoples in New England and Puritan English settlers there—incorporated European mirrors into their preexisting beliefs about and uses for reflective objects and made meaning with reflective devices as they circulated among, and between, their populations. Chapter 3 considers the evidence by which we can chart both interest in and ownership of mirrors from the earliest era of colonization through the nineteenth century among people of African, Native, and European descent. The evidence for African American ownership of mirrors is thin, given the societal failure to document and preserve what people of African descent owned. What can be done here is to look for evidence of interest in reflection (in both Africa and North America), consider archaeological evidence, and seek out textual references to mirror use and ownership. Native American trade records, as well as archaeological and documentary materials, provide evidence of ownership and use across time from different regions. White patterns of mirror ownership and use are explored through individual probate inventories and store records and include an extended discussion of how mirrors became a common household item over the course of the eighteenth and nineteenth centuries.

Chapters 4, 5, and 6 focus primarily on nineteenth-century North America and the mirror's role in fashioning both personal and collective identities. Chapter 4 examines how white Americans developed a complex, gendered relationship with looking glasses both as trusted sources of accurate visual information about the self and as sites at which their belief in the reliability of the eyes was increasingly tested. Their belief in vision as the "noblest sense" was tested when scientific evidence revealed, in the nineteenth century, that human vision did not always provide accurate information about the world. Despite their growing doubts about whether what could be seen could be trusted, whites strongly claimed a belief in the mirror as a reliable tool of visual mastery when they sought to use it as an object that could mark differences between themselves and those they sought to disempower in North America. Chapter 5 argues that whites sought out difference in Native American and African American mirror use to confirm their belief in white superiority and shows how the mirror solidified the racial categories that whites created and sought to enforce for themselves, as well as for people of African and Native descent. Chapter 6 takes the mirror and its meaning out of white hands and seeks to understand the incorporation of mirrors into nineteenth century African and Native societies in North America through the perspectives of these owners and users of this object, exploring the mirror's incorporation into ritual practices. Finally, the epilogue provides a glimpse from the close of the nineteenth century at how African American and Native American men and white women challenged white men's claims to define not only the mirror's meaning but also, by implication, the right to shape their own identities for themselves.

Fragmentary histories from three different cultures tantalize us with how early Americans—people of Native, European, and African descent; men and women; rich and poor; enslaved and free—interacted with mirrors. In attempting to understand the ways in which early American men and women interacted with reflective technologies, I try to recover something evanescent from the historical record. While encounters with looking glasses were frequent—albeit often fleeting—the moments when men and women translated their experiences, or even just referenced them, in textual material, were rare. Moreover, many of the men and women considered in this study would not have been able to leave a textual record even if they had wanted to do so. Thus, I have gathered evidence from a wide range of time periods, geographic regions, and kinds of sources in an attempt to understand the complex ways mirrors became an integral part of early Americans' lives. *In the Looking Glass* brings these fragments together to ask how mirrors shaped identity in early America from the time of the earliest European contact through the nineteenth century, exploring the various, multiple, and sometimes competing meanings early Americans attached to mirrors. It is my hope that readers will find the payoff worthwhile in the story I am able to tell about mirrors, meaning making, and identity formation among, and between, early Americans.

The Evolving Technology of the Looking Glass

In his *Memoirs of Louis XIV and the Regency* the Duke of Saint-Simon described the French Countess of Fiesque's acquisition of an early accurately reflective looking glass. Saint-Simon wrote, "When those beautiful mirrors were first introduced [the Countess of Fiesque] obtained one, although they were then very dear and very rare. 'Ah, Countess!' said her friends, 'where did you find that?'" The countess told them that she "had a miserable piece of land, which only yielded me corn; I have sold it, and I have this mirror instead. Is it not excellent? Who would hesitate between corn and this beautiful mirror?"[1] The expensive mirror the countess had purchased was an accurately reflective looking glass, the reflective technology that began to be widely produced in the early sixteenth century by Venetian craftsmen who perfected the production of both clear, flat glass and a highly reflective coating made from an amalgam of tin and mercury (i.e., quicksilver). Together, this glass and coating produced a mirror that surpassed the reflective capacities of earlier metal and glass mirrors. For at least a century after their introduction, however, these accurately reflective looking glasses remained exorbitantly expensive and could be obtained only by wealthy members of European society, largely as a result of the Venetian craftsmen's monopoly on their production. To lower the price of looking glasses, France and England sought to indigenize looking-glass production. Success over the course of the seventeenth century made looking glasses in France and England more available and affordable.[2] Studies have found that after 1650, two-thirds of Parisian inventories included a mirror, and 88 percent of a sample of wealthy Londoners owned a looking glass in 1675, whereas mirror ownership among the "middling ranks" across England rose from 22 percent in 1675 to 40 percent by 1725.[3]

While human beings have always had some opportunities to glimpse their own reflections using water and naturally occurring minerals with reflective properties, manmade reflective technologies underwent dramatic advances in the early modern era. To understand the import of these changes, we must begin long before the introduction of those prized accurately reflective looking glasses that were all the rage in the wealthiest European households in the seventeenth century. Understanding how earlier metal and glass mirrors often produced a dark and murky reflection enables us to see how reflection was linked to ritual and magic, as will be discussed in chapter 2, in which practitioners frequently claimed to be able to see things in mirrors that were not physically present before them or to find a connection between this world and the world beyond in those veiled reflections. Moreover, understanding how both early and later mirrors remained in circulation together over the course of two centuries helps explain why, even as the accuracy of looking glasses improved, later versions of this technology did not easily shed the preexisting associations with ritual and magic. Not surprisingly, language also adapted as mirror technologies underwent these changes. Exploring both evolving mirror forms and language allows us to trace the movement of different kinds of mirrors into North America and observe the varied uses for this object.

From Metal to Glass Mirrors

Since ancient times, Europeans had produced metal objects with reflective qualities. When craftsmen coated metal objects in molten tin to prevent rusting, they also produced a finished object with a reflective surface. This ancient practice was supplanted in the fourteenth century by tin-plating in which craftsmen coated thin sheets of iron or steel with tin before shaping the metal sheets into objects.[4] Both tinning and tin-plating produced a sheen on finished objects, including metal mirrors, similar to silver. The Englishman F. W. Gibbs, who studied the rise of the tin-plate industry in his native land, observed that "we no longer regard the finish of tinplate as in any way a substitute for silver, yet this characteristic of hand-beaten ware was repeatedly emphasized throughout the seventeenth and eighteenth centuries." Others who commented on the tin-plating industry also variously described tinplate as having a "lustre" or as "shining."[5] Other metal mirrors, called "speculum" or "steel," were composed of a highly polished copper and tin alloy but were significantly heavier (and thus less attractive for shipping overseas and across North America) than those made by tin-plating.[6]

During the twelfth century, European craftsmen also began producing glass mirrors, which became increasingly common from the thirteenth century

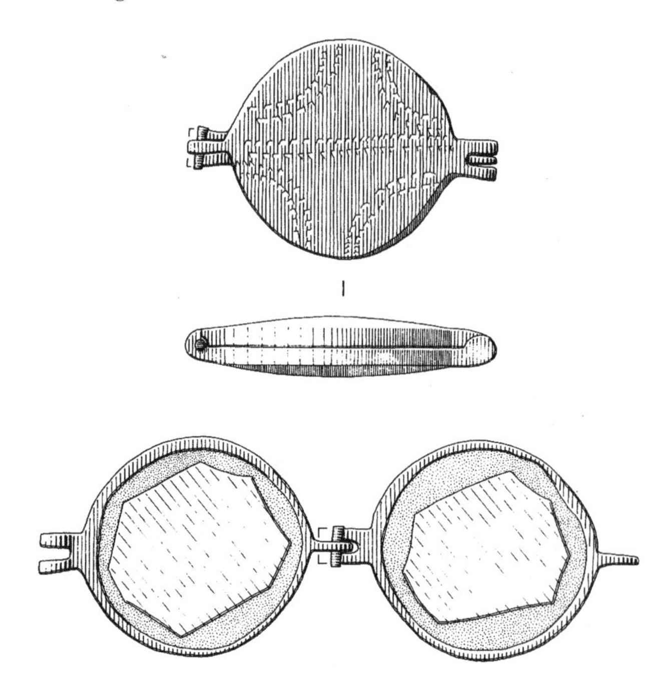

Figure 1.1. Fourteenth-century glass mirror, Winchester, England. From Martin Biddle, *Object and Economy in Medieval Winchester*, pt. 2 (Oxford: Oxford University Press, 1990), 655–56. Cat. no. 2103, fig. 178. Courtesy of Oxford University Press and the Winchester Excavations Committee.

onward. These mirrors were made by blowing a glass sphere and then coating its interior with a reflective mixture of molten metal. Once coated and cooled, the globe was cut into small, convex circular pieces. These small, fragile, glass mirrors could be enclosed by cases made of bone, horn, ivory, metal, or wood, which might be plain or carved.[7] The earliest known documentary evidence of these mirrors in England comes from 1371. That year, two cargoes destined for London from Bruges contained "two tuns" of mirrors; in 1384 another one thousand mirrors arrived in England. Twenty-six medieval glass mirrors in cases have been recovered archaeologically in the United Kingdom. Four fourteenth-century examples came from Winchester, Hampshire. One of these copper-alloy cases, when closed, is circular with a protrusion on both sides: on one side the hinge, on the other the clasp. When open, the mirror case is 82 mm across (a little more than three inches). This small mirror case had two pieces of convex, irregularly shaped mirror glass in it—one on each side of the case (fig. 1.1).[8]

What kind of reflectivity did these early glass and metal mirrors offer to observers hoping to catch their reflection in them? All of these mirrors would have been quite small—most of them would probably have fit in the palm of a person's hand. The glass mirrors had a convex reflective surface, causing a distortion in reflection; additionally, the glass often had a greenish hue and might be marred by small bubbles.[9] These attributes—convex shape, greenish cast, and bubbles—all impeded the ability of early glass mirrors to reflect an image accurately. Yet the quality of these early glass mirrors varied greatly: some undoubtedly "illustrated little more than the notion of reflection," whereas others allowed viewers to see themselves clearly in a "well-defined image."[10] Metal mirrors had at least one advantage over convex glass mirrors: their flatness, which did not distort an observer's mirror self as convex mirrors did. Seventeenth-century French North American fur trader Peter Radisson described the metal mirrors he stocked as sufficient for observers to "admire themselves."[11] People gazing into these early glass and metal mirrors may have been quite satisfied with their reflective capacity in a way someone several centuries later (and accustomed to peering into an accurately reflective looking glass) would not have been.

Although metal mirrors and convex looking glasses would remain in circulation through at least the early eighteenth century in Europe and North America, both were eventually eclipsed by accurately reflective looking glasses. The development of these accurately reflective looking glasses is generally credited to early sixteenth-century Venetian craftsmen on the Isle of Murano. Documentary evidence does show trade orders for flat glass and quicksilver from Artois, France, as early as 1312. This order significantly predates the famous Venetian craftsmen, although the Venetians were certainly the ones who popularized this type of looking glass and gained a monopoly on its production at the beginning of the sixteenth century.[12]

Making these accurately reflective looking glasses required highly skilled craftsmen. These artisans first had to produce a piece of flat, clear glass. Before the late 1680s, this glass plate, known as broad or cylinder glass, was produced by blowing glass. In this process a craftsman used an iron pipe to blow a "long, tubular bubble" out of molten glass. The ends of the bubble were cut off and it was opened on one side. Then, "as the glass was heated it was encouraged to open out along the cut until it lay flat on the plate" (fig. 1.2). These glass plates were limited by the size of the glass bubble to "about forty-five inches in length and thirty inches in width" but could be cut into a wider

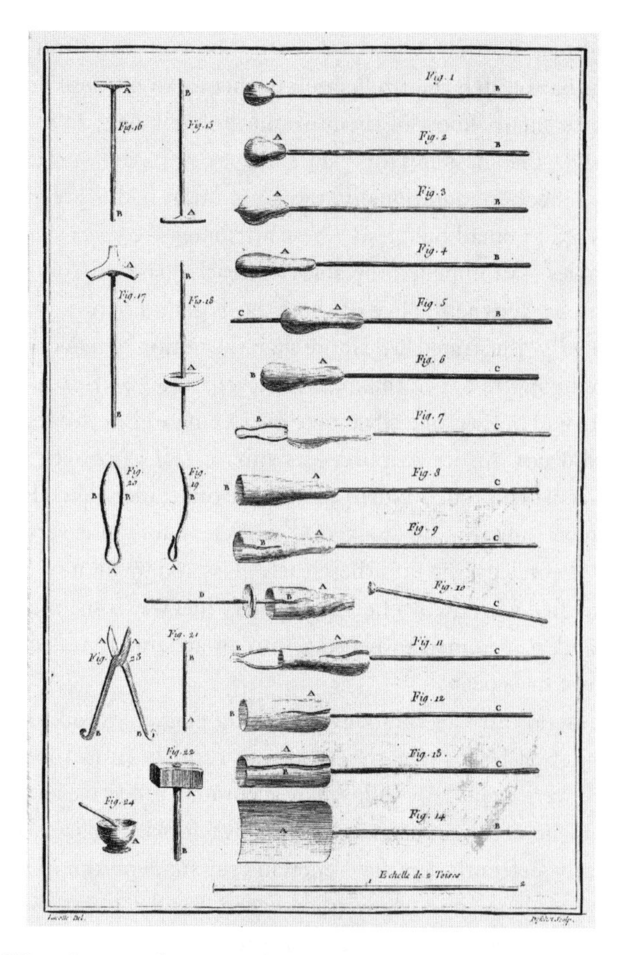

Figure 1.2. "Manufacture of Mirror Glass—Hand-Blown Glass." Plate 38: *Hand Blown Glass, Series of Operations and Tools.* From *The Encyclopedia of Diderot & d'Alembert Collaborative Translation Project* (Ann Arbor: Michigan Publishing, University of Michigan Library, 2010), http://hdl.handle.net/2027/spo.did2222.0001.464. Originally published as "Manufacture des glaces—Des glaces soufflées." Plate XXXVIII: *Glaces soufflées, opérations progressives et outils.* From *Encyclopédie ou dictionnaire raisonné des sciences, des arts et des métiers*, vol. 4 (plates) (Paris, 1765). Courtesy of the ARTFL Encyclopédie Project, University of Chicago.

variety of shapes than circular convex glass mirrors. Blown glass could never be completely flat, however, because "the inner and outer surfaces of the cylinder were not equal in length, and thus, when flattened, slight undulations called 'cockles' or 'waves' were produced."[13]

The zenith of blown-glass production was reached at the Hall of Mirrors in Versailles, constructed between 1678 and 1684. Each of the seventeen mirror-filled arches at Versailles contains twenty-one individually blown glass mirrors. But if construction at Versailles had begun only a few years later, it is likely that a new kind of glass, produced beginning in 1688, would have been used in the project. Casting glass involved pouring molten glass into flat metal plates (fig. 1.3). Casting quickly became the preferred method because it produced clearer glass as well as larger and thicker plates that were less susceptible to breaking during production.[14]

After the plate glass had been ground and polished to smooth its surface, craftsmen applied the reflective coating. Applying the tin-mercury amalgam was a complex, time-consuming process, as *A Dictionary of Arts and Manufacture* (1858) described it: the craftsman first prepared the marble table on which the glass plate would be placed. He "sweeps and wipes its surface with the greatest care, along the whole surface to be occupied by the mirror-plate; then taking a sheet of tin-foil adapted to his purpose, he spreads it on the table, and applies it closely with a brush, which removes any folds or wrinkles." Then he applied quicksilver "so that the tin-foil is penetrated and apparently dissolved by the mercury." More mercury was added "to form everywhere a layer about the thickness of a crown piece." The glass plate was then placed on the table under "a great many weights, which are left upon it for twenty-four hours, under a gradually increased inclination of the table." Finally, after being moved to a wooden table, "the mirror has its slope graduated from day to day, till it finally arrives at a vertical position." It could take up to a month for all of the excess mercury to drain off the mirror.[15]

Looking-glass plates were manufactured exclusively in Europe until the middle of the nineteenth century. In 1826 John Doggett, "one of the leading American looking-glass makers and importers of glass plates," wrote to his senator to ask that the duty on unsilvered glass plates be lowered. Doggett believed that it would "give some encouragement for Silvering here, and would not interfere with any Manufactory as there is none in the Country." By 1850, however, the American glass industry began attempting to produce plate glass; after 1884 the Pittsburgh Plate Glass Company made plate glass at a profit.[16]

At the end of the nineteenth century a new reflective coating for looking glasses came into wide use. This new reflective coating, developed in 1835 by German Justus von Liebig, used a less "poisonous and dangerous" silvering

Figure 1.3. "Manufacture of Mirror Glass—Cast Glass." Plate 24: *Mirror Glass, Operation of Pouring and Rolling.* From *The Encyclopedia of Diderot & d'Alembert Collaborative Translation Project* (Ann Arbor: Michigan Publishing, University of Michigan Library, 2010), http://hdl.handle.net/2027/spo.did2222.0001.463. Originally published as "Manufacture des glaces—Des glaces coulés." Plate XXIV: *Glaces, l'opération de verser et de rouler.* From *Encyclopédie ou dictionnaire raisonné des sciences, des arts et des métiers*, vol. 4 (plates) (Paris, 1765). Courtesy of the ARTFL Encyclopédie Project, University of Chicago.

process. It produced "a more brilliant and hard-looking reflecting surface" than quicksilver.[17] This new silvering process did not come into regular use in mirror production until the end of the nineteenth century, however, so the accurately reflective mirrors considered in this study used quicksilver to create their reflective surfaces.

Although the glass in American looking glasses was made in Europe, by the end of the eighteenth century American craftsmen frequently completed the construction of American looking glasses by framing the glass. American craftsmen did not widely practice the framing of imported looking-glass plates earlier because the labor costs to do this work were cheaper in London. By the 1790s, however, according to one study, it was much more common in Philadelphia to see "tradesmen [who] advertised looking glass plates for sale imported from England, France, and Italy" that were to be framed in the United States. In Philadelphia, "the last specific mention . . . of looking glasses imported from England or Europe appeared in an advertisement of John M'Elwee dated July 10, 1795." After this, Philadelphia dealers sold only imported looking-glass plates that American artisans had framed. Elsewhere, both American-framed looking glasses and European-framed ones continued to be sold into the nineteenth century. In 1818, for example, Francis Faverie, a carver and gilder in Richmond, Virginia, advertised that he had in stock both six looking glasses "framed in the newest fashion, with Sceptre Pillars, the Plates of English manufacture" and "9 Mahogany Framed Glasses, English, 26 by 18 inches."[18] Over time, however, looking glasses were much more likely to have been framed in North America.

The Language and Forms of Mirrors

During these centuries of transition from less-accurate reflective devices to accurately reflective looking glasses, the language used to describe mirrors evolved, as did the available forms of mirrors. The term *looking glass* emerged to differentiate glass mirrors from their metal counterparts. During the two centuries (between ca. 1500 and ca. 1700) when metal mirrors, convex glass mirrors, and accurately reflective looking glasses were being produced and used simultaneously, there was not, however, an absolute demarcation of the term *looking glass* from their metal counterparts, which were known as "mirrors."[19] Generally, *mirror* referred to metal and *looking glass* to glass mirrors. But, as in the examples of seventeenth-century North American fur traders John Pynchon and Peter Radisson, a "tin glass" and "looking-glasses made of tine [*sic*]" referred to metal mirrors.[20]

Sometimes, clues beyond just the object's name can be used to determine an object's reflective surface in these early records. Small, tin-plated metal mirrors became a popular item in the trade with Native Americans in New England and Hudson's Bay during the seventeenth century. The earliest examples of the term *mirror* in North America likely described these tin-plated metal mirrors. In 1524, Italian explorer Giovanni da Verrazano's expedition explored the East Coast of North America. Entering what is now Narragansett Bay, Massachusetts, Verrazano recounted how "boats full of people" greeted them. As the boats approached, Verrazano's men threw the gathered Narragansett "a few little bells and mirrors and many trinkets, which they took and looked at, laughing, and then they confidently came on board ship." That the "mirrors" were made of metal is suggested by how the Europeans handled them: tossing them into the natives' boats suggests they were not fragile. Along the California coast in 1602, Spanish explorers distributed "beads of colored glass, artificial garnets, hawks' bells, mirrors, knives, cheap scissors, Parisian tops, and some articles of clothing." A clue here that these Spanish explorers might have brought metal mirrors is that the item is listed with other metal items: hawks' bells, knives, and scissors rather than with the items made of glass (which also appear grouped together in the listing).[21]

By the mid-sixteenth century, the term *looking glass* was in use by European explorers to North America. Heading north from Florida in 1562, French explorer Jean Ribault gave one Native person he encountered "some little bracelets of tin, silver gilt, a sickle, a looking-glass, and some knives." And while waiting for the spring thaw to free their ship in 1611, Englishman Henry Hudson's expedition, stuck at the southern end of the Canadian bay that now bears his name, gave a local resident "a knife, a looking-glass, and buttons."[22] Unfortunately, trade records and early explorers' accounts do not reveal whether a glass mirror was convex or flat. We know from the archaeological recovery of the cargo of *La Belle*, one of the ships in French explorer Robert de la Salle's 1686 expedition to establish a colony at the mouth of the Mississippi, that convex glass mirrors were still arriving in North America near the end of the seventeenth century. *La Belle*'s cargo contained sixty convex glass mirrors coated with lead, housed in "tin containers that measured about two inches in diameter and about 3/4 inch in thickness, with lids slightly larger in diameter and about 3/8 inch deep."[23] The likelihood that glass mirrors were flat—and therefore accurately reflective—increased over time as the price of these mirrors decreased.

As the term *looking glass* entered the language during the sixteenth century, it was occasionally used for at least one other object that shared the mirror's ability to help humans see something clearly. *Looking glass* was sometimes used in sixteenth- and seventeenth-century England and North America as slang to describe a urinal or chamber pot. Clear glass urinals were used by doctors at this time to search a patient's urine for signs of disease. As poet John Collop wrote in his poem of the mid-1650s "A Piss-Pot Prophet": "Hence looking-glasses, Chamber pots we call, / 'Cause in your pisse we can discover all."[24]

By the eighteenth century, however, *looking glass* was the standard term used to describe only accurately reflective glass mirrors, which had by then almost entirely eclipsed metal or convex glass mirrors. And *looking glass* had replaced *mirror* as the common term for this object. One study, for example, found that in probate inventories from rural towns in Suffolk County, Massachusetts, the term *looking glass* appeared 153 times, while *mirror* was entirely absent from the descriptions of this object in 109 inventories taken between 1675 and 1775. By the end of the nineteenth century, however, the words had become reversed: *mirror* had become the common term (and remains so to this day), whereas *looking glass* had fallen into disuse.[25] To provide variety in this study, I do use "accurately reflective looking glass," "looking glass," and "mirror" interchangeably to describe accurately reflective looking glasses, while noting whenever a mirror may have been made of metal or of the earlier, less accurately reflective glass.

Accurately reflective looking glasses served a variety of purposes in early America. Many mirrors—especially those acquired by people of more modest means—were very small. An example of such a mirror is held in the collection of the Winterthur Museum, Garden and Library. This small (6 in. × 5 in.) looking glass was made sometime during the first two decades of the eighteenth century in Europe and imported to the colony of Massachusetts (fig. 1.4). The easiest reflection that could be accessed using such a small mirror would be a person's face. The Winterthur mirror may have been similar to a small mirror listed in a New York probate record in 1728, as part of the estate of New York colonist William Parsell. At the time of his death, Parsell was a slave owner who held in captivity an African man, Harre; two African boys named Dick and Tobe; an African girl, Doon; a Native woman named Isabel; and a Native boy, Tom. We can imagine the possibility that this one small mirror might have provided access to an image of the faces of men, women, and children of European, Native, and African descent, both enslaved and free.[26]

Figure 1.4. Miniature looking glass, 1700–1718, Boston, Massachusetts; pine, glass, paint, metal. Museum purchase, 1955.103.6. Courtesy of Winterthur Museum.

Although looking glasses of any size could be used to access a reflection of the self (or at least a portion thereof), larger looking glasses also served architectural, decorative, and design purposes, especially in the homes of wealthy Americans. In homes illuminated only by candlelight, mirrors were thought to increase the amount of light in a room by reflecting and multiplying that light.[27] One popular form that sought to maximize the increased light produced by a mirror and proximate candles was the sconce glass: mirrors to which candleholders had been attached (fig. 1.5). Increasingly after 1688, when larger glass could be produced, mirrors became grand pieces of furniture in wealthy homes and were used to increase available light. As Elizabeth Donaghy Garrett argues in *At Home: The American Family, 1750–1870*, "the dramatic alliance of artificial light and plate glass was clearly stated in the conspicuous use of sconce glasses in parlor, dining room, and bedchamber and was reiterated in the strategic placement of candles, lamps, and chandeliers before all manners of looking glasses."[28]

Figure 1.5. Sconce, 1710–40, England; glass, wood, gesso, brass. Bequest of Henry Francis Du Pont, 1959.184.1. Courtesy of Winterthur Museum.

In eighteenth- and early nineteenth-century interiors, large looking glasses were frequently placed over fireplace mantels (called chimney glasses) or between two windows (called pier glasses).[29] This arrangement made the room appear larger and maximized the available light sources. The circa 1828 painting *York, Pennsylvania, Family with Servant* features both a chimney and a pier glass (fig. 1.6). Furniture designer Thomas Sheraton's early nineteenth-century *Cabinet Dictionary* emphasized the importance of both mirrors and light in a well-appointed room: "The furniture used in a drawing-room are sofas, chairs to match, a commode, pier tables, elegant fire-screens, large glasses, figures with lights in their hands, and bronzes with lights on the cap of the chimney piece, or on the pier tables and commodes." As Sheraton reported, pier glasses

Figure 1.6. York, Pennsylvania, Family with Servant, c. 1828, American; oil on panel; 15 3/4×11 3/4 in. Saint Louis Art Museum. Bequest of Edgar William and Bernice Chrysler Garbisch, no. 24:1981.

were available in a variety of sizes from the standard 16×28 in. glass up to a 75×177 in. custom-order glass.[30]

New mirror forms emerged by the end of the eighteenth century. One was a circular, convex glass called a "mirror." This mirror was intended as a decorative feature that produced what were considered "agreeable" reflections (i.e., distortions) of the rooms in which they were placed. Popular in France as "girandole mirrors," they had a "candelabra fixed before" them; when of English manufacture, these mirrors usually did not have a candelabra (fig. 1.7). The reflections they produced were "compressed, delicately distorted . . . accompanied by the

Figure 1.7. Girandole mirror, 1800–1810, England; wood, gesso, gilt, glass, paint. Bequest of Henry Francis du Pont, 1957.523.1. Courtesy of Winterthur Museum.

intensification of colors, lights and shadows, [and] provided a delightfully different reflection."[31] In bedrooms, framed full-length mirrors, called cheval glasses, which could be tilted to the viewer's desired angle, became popular at this time as well, joining dressing-table glasses and shaving stands as sources of reflections of the self and of light in a home's private spaces (fig. 1.8).[32]

Technological innovations made the most accurate looking glasses that had ever been mass produced available in a wide variety of shapes, sizes, and styles in early America. The changes in mirror making that brought about this development created multifarious mirror types and words used to describe this object between ca. 1500 and 1700, leaving a sometimes confusing historical record for this period. Nonetheless, there is much to be learned about how people apprehended their reflections and understood this object during that time period, which I will explore in the following chapter. By 1700, accurately

Figure 1.8. Cheval Glass, ca. 1815. Made in New York. Mahogany, mahogany veneer, brass, 75×44 7/8×27 3/4 in. (190.5×114×70.5 cm). Gift of Ginsburg and Levy, Inc., in memory of John Ginsburg and Isaac Levy, 1969 (69.183). The Metropolitan Museum of Art, New York, NY, USA. © The Metropolitan Museum of Art. Image Source: Art Resource, NY.

reflective glass mirrors were well established as the primary type of mirror, and they were most often known as looking glasses. In the eighteenth and nineteenth centuries, mirrors became important pieces of interior decoration and furniture as well, serving a range of aesthetic purposes. Moreover, these objects—from the plain to the most opulent—began to play an integral role in people's self-fashioning.

CHAPTER TWO

First Glimpses

Mirrors in Seventeenth-Century New England

Puritan colonist Mary Rowlandson's *The Sovereignty and Goodness of God* (1682) described her three-month captivity during Metacom's War, the 1675–76 conflict between Native peoples and English settlers in New England. Mary, her husband Joseph, and their three children lived in Lancaster, Massachusetts, where Joseph served as the town's Puritan minister. Native raiders attacked Lancaster in February of 1676 and took her and the children into captivity. Quinnapin, a high-ranking Narragansett, purchased Rowlandson and adopted her into his household. She wrote *The Sovereignty and Goodness of God* to recount her experiences and to portray, as historian Laurel Thatcher Ulrich describes it, "a cosmic battle between foul heathens and fair Christians," in which Rowlandson believed herself to have participated. Several events that took place during her captivity, however, complicated this intended motif. One occurred near the end of the captivity, after Quinnapin and Rowlandson were separated for three weeks. Seeing her for the first time after this absence, Quinnapin looked askance at Rowlandson's physical appearance and asked her when she had last washed. Rowlandson told him that she had not washed in some time. In response, Rowlandson wrote, Quinnapin "fetcht me some water himself, and bid me wash, and gave me the Glass to see how I lookt; and bid his *Squaw* give me something to eat." Quinnapin's concern for Rowlandson's physical condition prompted him to offer her what he deemed to be the necessary items to remedy her state: water, nourishment, and a mirror. Water to wash and food to eat need little explanation, but what of the mirror? Why did the Narragansett Quinnapin, the "foul heathen," have this item of European material culture to offer his "fair Christian" captive? Under what circumstances might Quinnapin have acquired this mirror, and why would it have been an item he desired, kept with him, and was willing to share with this woman? What meaning might this mirror have held, as well, for the Puritan Mary

Rowlandson, as she stretched out her hand to accept it from Quinnapin and cast her eyes upon her reflection?[1]

The first known introduction of European mirrors into Narragansett society had happened more than 150 years before Quinnapin offered his mirror to Mary Rowlandson in 1676. These first mirrors were likely metal ones that Italian explorer Giovanni da Verrazano's expedition brought with them and threw overboard to a group of Narragansetts who came to greet the European ships in 1524. Verrazano observed that the Narragansett were only interested in acquiring some of the items the Europeans had brought, especially "little bells, blue crystals, and other trinkets to put in the ear or around the neck." In contrast, "they did not appreciate cloth of silk and gold, nor even of any other kind, nor did they care to have them; the same was true for metals like steel and iron, for many times when we showed them some of our arms, they did not admire them, nor ask for them, but merely examined the workmanship. They did the same with mirrors; they would look at them quickly, and then refuse them, laughing." The items rejected by the Narragansett in their encounter with Verrazano's expedition were ones that would become desirable to them later. The Narragansett rejection of European mirrors and other goods—while expressing interest in obtaining some European wares—reveals the complex human process of negotiation and assessment Native peoples made in their first encounters with Europeans.[2]

By the early seventeenth century, the era of exploration had given way to the initial period of European settlements in North America. Trade records between these Europeans and Native peoples reveal significant information about the kinds of goods Native peoples had access to through the fur trade. Trade records are not a perfect proxy for consumer interest, but it behooved traders to learn what goods were desired by their Native trading partners. Native peoples in North America regularly expressed their preferences for particular items and let their dissatisfaction be known when delivered goods failed to meet their expectations. Traders adjusted their merchandise accordingly.[3]

Seventeenth-century North American trade records reveal that Native peoples initially sought out smaller items like those "little bells, blue crystals, and other trinkets to put in the ear or around the neck" that garnered interest during Verrazano's 1524 expedition. Quickly, however, Native peoples came to prioritize cloth and metal goods over these smaller items. In 1623 Governor William Bradford of Plymouth Colony (settled by the English in 1620) observed that the local people did not want "toyes and trifles" but sought "good and substantial commodities," including kettles, hatchets, and cloth. John Lederer, an early explorer in North Carolina and Virginia's interior territory in 1669 and 1670, observed this distinction as well. Of the "neighbour Indians"—those living nearest the Europeans—Lederer wrote of a desire specifically for cloth and a wide variety of "edg'd tools," as well as much negotiating over price. But of "remoter Indians"—those who had less exposure to Europeans and their merchandise—he observed that they "admire such trinkets" as "small Looking-glasses, Pictures, Beads and Bracelets of glass, Knives, Sizars, and all manner of gaudy toys and knacks for children," which they "will purchase . . . at any rate." This shift in priority over time did not mean, however, that Native peoples stopped trading for some of the smaller items—mirrors included—only that, as archaeologist Bruce Trigger put it, these items were not "at the top of their shopping lists."[4]

Seventeenth-century New England traders stocked robust supplies of both metal and glass mirrors for Native peoples. The earliest known European mirrors intended for Native ownership appear on William Pynchon's bill to John Winthrop Jr. at coastal Roxbury, Massachusetts, dated March 17, 1635. Both men were early settlers in New England and active in the fur trade. William Pynchon arrived in 1630 and was involved in the fur trade by 1631. Winthrop, who had arrived in the Massachusetts Bay Colony in 1631, was preparing to plant the English settlement of Saybrook at the mouth of the Connecticut River. On this particular day Winthrop purchased a variety of objects that Europeans regularly proffered Native peoples in the fur trade: fifteen pieces of cloth, two dozen looking glasses, three quart pots, four dozen jaw harps, four dozen steel awl blades, one dozen shallow bowls with handles (called porringers), one dozen metal spoons, and fifteen hoes. Winthrop likely used these goods as part of the process of acquiring land for the town of Saybrook.[5]

English settlers offered similar collections of goods to Native peoples in exchange for land as they established other settlements along sixty miles of the Connecticut coastline. In February of 1639 the English established Milford, acquiring land from Paugussett leader Ansantawae in exchange for "six coats, ten blankets, one kettle and a quantity of hoes, knives, hatchets and looking-glasses." That same year, the English enacted their founding of Guilford in part with an exchange of "twelve coats, twelve fathom of wampum, twelve looking-glasses, twelve pairs of shoes, twelve pairs of stockings, twelve hatchets, four kettles, twelve knives, twelve hats, twelve porringers, twelve spoons and two English coats." The pattern continued in 1640 and 1641 at Norwalk and Stamford, where the initial exchanges repeated the pattern of goods enumerated above, including looking glasses.[6]

These early coastal English settlements were located in the area scholars now refer to as "southern New England." Seventeenth-century southern New England included coastal Massachusetts, Connecticut, Rhode Island, and the eastern end of Long Island. The region was inhabited by Algonquian-speaking peoples, including the Narragansett, Pokanoket (also known as Wampanoag), Pequot-Mohegan, Niantic, Nipmuck, and Montauk, who, according to historical archaeologist William Turnbaugh, shared "close cultural and linguistic affinities" with one another.[7] Although the English settled first in these coastal locations now known as southern New England, some quickly moved inland to the fertile Connecticut River Valley, some one hundred miles west of the Atlantic coastline. William Pynchon moved his fur-trading enterprise from coastal Roxbury into the Connecticut River Valley, becoming one of the founders of the town of Springfield, Massachusetts, in 1636. An estimated five thousand Native people lived in the Connecticut River Valley around the time of European arrival there, in the communities and surrounding territory of Woronoco, Agawam, Norwottuck, Pocumtuck, and Squakheag near the Connecticut River.[8] Taken together, the Connecticut River Valley and southern New England offer a uniquely rich combination of trade records, narrative accounts, and archaeological excavations to draw on in our seeking to understand the incorporation of mirrors into the Native societies that made their homes in these regions.

In the Connecticut River Valley, John Pynchon, William's son, had taken over the trade from his father and was running a very successful business at midcentury. According to historical archaeologist Peter Thomas, John Pynchon had a "virtual monopoly" over the fur trade in western New England.[9] Pynchon also kept meticulous records. These records provide great detail about the number and kind of mirrors available in the Connecticut River Valley. In the decade between 1652 and 1661, John Pynchon stocked his network of traders with at least 1,534 mirrors: 1,200 metal mirrors and 334 made of glass.[10]

John Pynchon called the 1,200 small, tin-plated metal mirrors "tin glasses," but he did not record their shape or size. Pynchon used the word *tin* to distinguish these metal mirrors from the glass mirrors he called either "looking glasses" or simply "glasses."[11] Their unusual name—"tin glasses"—reminds us of the transition being made during this period from metal mirrors to glass mirrors.[12] The remaining 334 mirrors Pynchon supplied had reflective surfaces made of glass. It is not possible to discern whether these mirrors were the older, poorly reflective glass mirrors or accurately reflective looking glasses. Pynchon had 152 he described only as "looking glasses" (146) or "glasses" (6), but he gave no clue about their size, shape, or frame. Another 149 of the mirrors were identified by their protective cases, called "boxes," made of tin (36), brass (62), or wood (63).[13] The shape of these cases is not known, although contemporaneous trade records indicate that at least some were circular, a practical shape if they housed convex, circular glass.[14] They might have looked something like the Hampshire mirror from two centuries earlier (see fig. 1.1). We do know that some round glass mirrors found their way into this area during the mid-seventeenth century. Recovered archaeologically from the Squakheag Fort Hill Settlement (which overlooked the Connecticut River in southern New Hampshire), occupied between 1663 and 1664, were two pieces of round glass mirrors. These mirrors were small, with a diameter of 3.2 cm (not quite 1.5 in.), although nothing about the frame or case of these mirrors is known.[15]

Pynchon also supplied a small number of "book looking glasses." The twenty-one mirrors of this variety had leather covers creating a resemblance to their namesake. This kind of mirror also appears in other contemporaneous trade records as "leather looking glasses."[16] Book looking glasses must have been made from accurately reflective flat glass because these mirrors could be cut into squares or, more likely in the case of a "book looking glass," rectangles.

If Pynchon's traders knew their customers' desires well, then all or most of these mirrors would have made their way into Native hands during this period. Given the population estimate at first contact of five thousand individuals, Pynchon's stock of 1,534 mirrors would have meant a possible individual ownership rate of almost 31 percent in the Connecticut River Valley by the early 1660s. If ownership rates were counted by family units, then the rate would be significantly higher because one mirror could have been shared by several family members. For example, if family unit size is estimated at five per unit and mirrors were distributed evenly throughout the population, then there would have been 100 percent mirror ownership by family unit, with each family possessing an average of 1.5 mirrors.

Beyond this information about types of mirrors and possible ownership patterns, however, the Connecticut River Valley yields no evidence of how these mirrors might have been used or the meanings Native peoples ascribed to them. Southern New England, that coastal region located one hundred miles east of the Connecticut River Valley, provides a missing context that can help us understand how Native people incorporated mirrors into their practices and beliefs in seventeenth-century New England. What southern New England lacks, however, are records that reveal how many mirrors were available to Native consumers there. But only one hundred miles separated the Connecticut River Valley from southern New England, not very far for trade items to travel. Indeed, documentary and archaeological evidence indicates that both regions shared a common robust supply of mirrors. Moreover, the available evidence for southern New England provides important clues about how Native peoples in the region used and understood these mirrors.

Roger Williams's *A Key into the Language of America* (1643) includes the only extended discussion of how seventeenth-century Native people in southern New England used and understood mirrors. Williams is best known as the controversial Puritan minister who, after being exiled from Massachusetts Bay in 1635, founded Rhode Island. Williams arrived in New England in 1631, initially settling at Boston. Within a year he and his wife, Mary, had moved north to Plymouth Colony, where Williams "began to cultivate relationships with neighboring Wampanoag peoples and others who frequently visited the English plantation." Williams returned to the larger Puritan settlement at Massachusetts Bay in 1633, from which he would be exiled for nonconformist teachings two years later. After his exile Williams moved to Narragansett Bay, where he set up a trading post at Cocumscussoc (located near present-day Wickford Harbor, Rhode Island). As historical archaeologist Patricia E. Rubertone observes in her study of Roger Williams, "Cocumscussoc provided Williams an unparalleled vantage point from which to observe Narragansett lifeways more closely."[17]

In 1643 Williams penned his insightful interpretation of the language and culture of the Narragansetts, which archaeologist William Simmons describes as "the most complete record we have of Narragansett culture in the seventeenth century," noting that "others who wrote of the Narragansetts fell short of [Williams] in their interest and penetration into the subtleties of native life." More recently, Rubertone has attenuated these claims by exploring how "major portions of the world Williams attempted to portray remained silent (and invisible) to him." Rubertone shows how elements of Narragansett culture, including familial structure and ritual life, were absent from *A Key*, but she does acknowledge that Williams "came to possess linguistic competence and a familiarity with local knowledge."[18]

Williams mentioned mirrors twice in *A Key*. In the first instance he provided two Narragansett words for looking glasses—*kaukakíneamuck* and *pebenochichauquânick*—and then entered into an imagined conversation with his European audience, anticipating that they would wonder what the Narragansett "do with Glasses, having no beautie but a swarfish colour, and no dressing but nakednesse"?[19] Williams's answer revealed his sensibilities about a common humanity shared by all. The Narragansett interest in looking glasses could be explained, Williams argued, because vanity was a universal human characteristic: "pride appeares in any colour, and the meanest dresse."[20] But then Williams singled out Narragansett women. He had already hinted that he was thinking mostly about the women by framing his question around "beautie." But he made the connection explicit when he wrote "and besides generally the women paint their faces with all sorts of colours."[21] If his European audience was surprised by the Narragansett interest in looking glasses when Williams first mentioned it, they would have found little unusual in Williams's connection of looking glasses to women, for mirrors were by this time already strongly associated with women and femininity in European cultures. And Williams was right that Narragansett women did paint their faces. But men also painted their faces, and Williams knew this: a few pages later in *A Key*, he introduced a discussion of painting practices with this list:

1. They paint their Garments, &c.
2. The men paint their Faces in Warre.
3. Both Men and Women for pride, &c.[22]

Here Williams gave pride of place to the men's practice of face painting in war and included men and women in the category "for pride, &c." But earlier when Williams considered mirrors, his gendered understanding of the use of this object by women shaped his observations. There he mentioned only women.[23] Mirrors were, from the earliest era of colonization, difficult for Europeans to disentangle from their ethnocentric assumptions about what they meant and the work they did.

The second time looking glasses appear is in Williams's discussion of the Narragansett belief in a dual soul.[24] According to Williams, the Narragansett believed in a soul known as *cowwéwonck*, which "derived from *Cowwene* to sleep, because say they, it workes and operates when the body sleepes." There was also, Williams continued, *míchachunck*, "the soule, in a higher notion, which is of affinity, with a word signifying a looking glasse, or cleere resemblance, so that it hath its name from a cleere sight or discerning, which indeed seemes very well to suit with the nature of it."[25]

Williams observed two things in this passage about "the soule, in a higher notion." The first, near the end of his description of míchachunck, was that the Narragansett word for this soul "hath its name from a cleere sight or discerning." Here Williams's meaning is plain: the word the Narragansett used to describe this soul was structurally related to—"hath its name from"—words meaning "clear sight" or "discerning." But at the beginning of this passage, when Williams mentioned mirrors, he said only that míchachunck is "of affinity, with a word signifying a looking glasse, or cleere resemblance." "Of affinity" might indicate a structural similarity, but it also could mean a connection not rooted in the words themselves. That Williams made it so clear a moment later when he did note a structural similarity between míchachunck and "cleere sight or discerning" suggests that here he meant to emphasize something different by "of affinity." J. H. Trumbull, who edited Williams's *A Key* in 1866, also observed that he saw no structural similarity between the words for looking glass and the word for this soul, noting that "possibly, Mr. Williams was mistaken as to the affinity of this word with one 'signifying a looking glass.' "[26] Rather than linking *soul* with *looking glass*, Williams might have linked both *soul* and *looking glass* to clarity of vision. We can look to other Native languages for comparison, where there is evidence that Native peoples connected European looking glasses not with a word for soul but with the verb *to see*. Although distinct languages, Narragansett, Ojibwe, and Cree all belong to the Algonquian family of languages. In both Ojibwe and Cree the verb *to see* and *mirror* share the same root.[27]

Native peoples also incorporated mirrors into their preexisting practices of divination—the use of material objects to see something not physically present. Historian Christopher Miller and curator and archaeologist George Hamell argue that before colonization, Native peoples had access to "water and crystal," as well as "polished free-state metals, metallic-ore mosaics, muscovite mica, and perhaps water- or grease-slicked polished stone surfaces." They used the reflective qualities of these objects to practice divination.[28] Scattered post-contact evidence from across the continent suggests the practice of divination using reflections. A Jesuit missionary living among the Huron of the Great Lakes Region in the mid-seventeenth century observed medicine men using water in divination. The missionary described the Huron as believing that they could "see the hidden desires in the soul of the sick person" by looking "into a basin full of water, and say that they see various things pass over it, as over the surface of a mirror." The medicine men claimed they could see various images in the water that enabled them to identify those desires. In recounting his mid-eighteenth century travels from his native Pennsylvania to Lake Ontario, John Bartram similarly observed: "However, I find different nations have different ways of obtaining the pretended information. Some have a bowl of water, into which they often look, when their strength is almost exhausted, and their senses failing, to see whether the spirit is ready to answer their demands." In *Myths of the Iroquois* the nineteenth-century anthropologist Erminnie Smith (1836–86) described a medicine man practicing divination, "who by the use of a small kettle boiled roots or herbs, and by covering the head with a blanket and holding it over the kettle could see the image of an enemy who had bewitched either some one [*sic*] else or himself." Frank Speck, an early twentieth-century anthropologist, described the Naskapi—an Algonquian-speaking people whose homeland is in the far northeastern portion of Canada (roughly Eastern Quebec and Labrador)—who practiced "divination by gazing," which Speck wrote was a "widespread practice." The Naskapi would look at "a decorated object, or a mirror, or into a pool of water until, through concentration upon the subject upon which information is desired, an image" was made visible. These accounts suggest that Native peoples used naturally occurring and other early reflective technologies for divination and later incorporated European mirrors into those practices.[29]

Archaeological findings provide the final set of evidence about the uses and meanings of mirrors among Native peoples in seventeenth-century southern New England. Extensive archaeology done at burial sites in this region has confirmed that a variety of glass mirrors similar to those that appear in Pynchon's records from the Connecticut River Valley was making its way into this region.[30] Sometime during the late sixteenth or early seventeenth century—the era just before permanent European settlement began in New England—the Algonquian-speaking peoples of southern New England underwent a transformation in burial practices. An earlier practice of burying the dead in "isolated internments, as opposed to burial in cemeteries" containing "few, if any, associated grave goods" gave way, as permanent European settlement began, to burials in distinct graveyards with increasing numbers and kinds of grave goods.[31]

Four southern New England Native burial grounds were studied over the course of the twentieth century: West Ferry, 1620–60 (Conanicut Island,

Rhode Island); RI-1000, 1650–70 (North Kingstown, Rhode Island); Burr's Hill, 1650–75, with some grave goods from the early eighteenth century (Warren, Rhode Island); and Pantigo, 1660–1728 (Montauk, Long Island).[32] For these cemeteries we have detailed information about the grave goods that were buried with the deceased. William Turnbaugh analyzed the material culture of three of these sites and found an increase in the number and diversity of European goods over the course of the seventeenth century. He identified 123 types of artifacts: West Ferry has forty-six of the types; RI-1000 has sixty-eight; and Burr's Hill has eighty-five.[33]

The four southern New England Native burial grounds yielded three mirrors, one each at West Ferry, Burr's Hill, and RI-1000.[34] At West Ferry the mirror was a very small piece of reflective glass inset in a brass wire ring.[35] At RI-1000 the mirror was a "wooden mirror box."[36] At Burr's Hill the mirror was described as "pieces of mirror and leather/bark frame."[37] If the frame was leather, then it was in all likelihood a book looking glass. Three additional pieces of glass that were not identified as mirrors were recovered from these sites, but all of them, I believe, were pieces of European mirror glass. All three are small glass circles. Two from RI-1000 are described as convex with "fragments of metal, possibly a frame or holder, [that] adhere to the edges and one face of each specimen." The third, from Burr's Hill, was part of a circular iron box and could have been a "box glass" like those in Pynchon's inventories.[38]

Counting all six items discussed above at these three sites as mirrors (the three identified as mirrors in the original findings and the three I strongly suspect were also mirrors), we find one mirror at West Ferry, two at Burr's Hill, and three at RI-1000—suggesting an increase in the number of mirrors over time, as Turnbaugh's findings would predict. The pattern Turnbaugh identified of increased diversity of goods is more difficult to judge because each site had different types of mirrors (mirror glass in ring, wooden mirror box, convex glass mirrors, etc.), meaning there is no evidence to support an increase in diversity of types of mirrors over time as has been documented for other material objects found in these burials.

Mirrors appear too infrequently in the graves to provide clues about cultural logic behind their placement, but it is possible that mirrors were also part of a larger category of material goods placed in graves, all of which had the capacity to deflect light.[39] In a 1975 article, "Magic Stones and Shamans," published in the *Bulletin of the Massachusetts Archaeological Society*, the archaeologist William S. Fowler speculated—he called it a "good hypothetical guess"—

about the power of deflection for Native peoples. Fowler argued that in an earlier era, one that ended around AD 1000, some Native burials included stones with a striking appearance because of their "color, odd surface effects, reflective qualities, high finished surfaces, or on occasion, skillfully incised work denoting human esteem."[40] Fowler speculated that "shamans" placed these stones in the graves because they were needed "to drive away an evil spirit" by deflection because "burial rites included the objective, among others, of preventing an evil spirit from interfering with the dead one's journey to the other world."[41] Fowler noted one later grave, from around the time of European contact, that also included an example of these kinds of stones. In this case, however, the eight quartz crystals seem to have served a different purpose. They were located "inside a birch bark envelope container, apparently the personal property of the buried woman at whose feet it lay," which Fowler further speculated could mean that the quartz crystals held special significance to the deceased and were not items that needed to be exposed in the grave to repel a malicious spirit.[42]

To continue down Fowler's speculative path, we might consider the possibility that the belief about the power of deflection and the need to guard the dead against those who would interfere with their journey to the afterlife persisted during the centuries-long period before European colonization when grave goods were not placed with the deceased.[43] If so, what kind of evidence from these later burials—when grave goods again became a part of burial rites—might support such a belief? One possibility is to look for objects that would be well suited to deflect an image of whatever passed before them. One such object would be a metal spoon; the backside of the bowl makes a convex reflective—and therefore deflective—surface. A convex reflective surface causes light rays to diverge and would be much more appropriate for deflection than the other side of the spoon, the concave surface, which would bring light rays together. Metal spoons were one of the most common items at the southern New England burial grounds, with fifty-nine spoons recovered from thirty-nine burials at the West Ferry, RI-1000, Burr's Hill, and Pantigo cemeteries. In at least some cases, the bowls of the spoons were placed upside down, meaning that the convex reflective surface faced up. Some of the spoons were placed in the graves with no evidence of having been used, suggesting that they might have been reserved for a ritual purpose. The spoons also appear more frequently in children's graves than in the final resting places of adults, and it is possible, as others have suggested, that these youngest members were deemed to be in need of more protection as they undertook their journey after death.[44]

Mirrors would also have deflective properties, especially the early convex glass mirrors. Could mirrors have also been placed in the graves because of their deflective potential? Two of the objects I have called potential mirrors had convex reflective surfaces. Another two, the "mirror box" and the potential mirror that was a glass in a circular iron box, were both likely made of convex glass as well because fragile, circular convex glass mirrors were often housed in protective boxes. The book looking glass would have been made of flat glass and therefore less useful for deflection, especially if it had been enclosed in its leather case at the time of burial, which would have hidden the reflective surface. The final mirror, that small piece of glass set in a brass ring, presents at least two possible interpretations. Simmons noted that this ring, along with "an iron buckle, [and] three teardrop-shaped shell pendants" were "rusted by the buckle, and they may have been wrapped in a packet and thrown in as an afterthought."[45] If they were wrapped in a packet, then they would conform to Fowler's example of the potentially magical stones that were also contained by a packet and therefore not meant to deflect a malicious spirit. But, if they were not wrapped in a packet, then perhaps this ring was desirable for deflection. Such a small piece of mirrored glass would not have been useful for its reflective capacity, but it would have been able to catch and deflect light—to shimmer—and thus could have been thought of as a deflective surface.

Perhaps more significantly, all three of the graves with known mirrors also contained metal spoons, although only one of the graves with a potential mirror included a metal spoon. Assuming that all six of these glass items were actually mirrors, then two-thirds of the mirrors appear with spoons. The possibility of a faint pattern emerges here, suggesting that deflective powers may still have been desirable in burials during the period of European contact with Native peoples in southern New England and that both mirrors and spoons had the potential to serve as deflective devices in burials. But here we reach the limits of the evidence.

We can return now to that looking glass Quinnapin stretched out his hand to offer to Rowlandson during her captivity in 1676. Unfortunately, we cannot know what kind of mirror it was. Rowlandson refers to it only as a "glass." Whether it was a convex early glass mirror or an accurately reflective looking glass is impossible to ascertain. Given that both Quinnapin and Rowlandson seemed confident that it would allow Rowlandson to get a reasonably clear view of herself, we can conclude that its reflective capacity was relatively good.

Whatever its physical form may have been, that small trade mirror is now laden with possible meaning. For Quinnapin we can ask how these complex layers of its possible meaning—connected to European assumptions about the gendered uses of this object, the mirror's linguistic connection to clear sight, and its possible use in divination and for deflection—articulate with the evidence we can gain from an analysis of the encounter between Quinnapin and Mary Rowlandson. First, Quinnapin's ownership of this mirror explicitly challenges Roger Williams's assumption that only Narragansett women found looking glasses a desirable item to possess. Moreover, if Quinnapin believed in or had experienced the power of reflection for divination or deflection, those beliefs did not preclude his handing over this mirror to Mary Rowlandson, a woman from a culture against which his people were then at war, to use for a very practical purpose. In that moment, looking at Mary Rowlandson's dirty visage, Quinnapin clearly linked the mirror to the clear sight it would provide Rowlandson of her face and how it would enable her effectively to improve her appearance. He also may have been thinking about her impending return to her family and how he wanted her to appear as her captivity ended. Finally, Quinnapin's desire that Rowlandson use the mirror for this purpose is suggestive of how he might have used this object in his own life.

What of Mary Rowlandson? What did she think as she took the mirror from Quinnapin's hand that day? What did her culture and faith tell her about the meaning and use of mirrors? Rowlandson was born around 1637 to John and Joan White in Somerset County, England. She moved with her family to the Massachusetts Bay Colony in 1639 and settled in the coastal town of Salem. John White, Mary's father, was one of the founders of Wenham, six miles north of Salem. In 1653 the family moved fifty miles farther inland to Lancaster, Massachusetts. It was there that Mary would meet her future husband, Lancaster's Puritan minister, Joseph Rowlandson. They married by 1656 and had four children, three of whom survived infancy. It was from Lancaster that Rowlandson and her three children were taken in February of 1676 during Metacom's War. After the captivity, Joseph Rowlandson moved his wife and their two remaining children (daughter Sarah had died during the captivity) to Wethersfield, Connecticut, along the Connecticut River in the Connecticut River Valley, where he died in 1678. When Rowlandson penned her narrative in the years after the captivity, she did so surrounded by material comfort. As the historian Neal Salisbury recounts it, after the captivity, Mary's husband, Joseph Rowlandson, parlayed the fame surrounding Mary's experience into a position at the church at Wethersfield where he became "one of the highest paid clergymen in New England." The year after Joseph Rowlandson's death, Mary Rowlandson married another wealthy man in Wethersfield, Samuel Talcott.[46]

Reflective objects had a long history of being associated with magic and ritual in Rowlandson's birthplace of England. Early reflective devices in England included water, metal mirrors, poorly reflective glass mirrors, and polished stones. In medieval England, objects with reflective properties were important tools used by those who practiced divination. In 1287 and again in 1311, Church of England bishops decried magicians and named tools linked to their roles as conjurers and diviners, including the reflective surfaces of sword blades, water, metal mirrors, stones, and even fingernails.[47] John Dee, the sixteenth-century English scholar of mathematics, astrology, and medicine, used an obsidian mirror taken from the Mexica in Central America during the European conquest in his magic and crystallomancy. Dee practiced his craft before Queen Elizabeth I, who, in 1575, visited Dee and "did see some of the properties of that glass, to her Majestie's great contentment and delight." Now in the collection of the British Museum, John Dee's obsidian mirror produces a remarkably clear reflection, yet the darkness of its surface invites speculation.[48] Countless others, less well known than Dee or lost altogether to obscurity, practiced magic using mirrors and other reflective surfaces during the Elizabethan Era as well.[49]

Reflections could also be harnessed by people seeking to root out and eliminate those who practiced magic among them. The English Puritan minister Richard Bernard reported in his 1627 *Guide to Grand-Jury Men . . . in Cases of Witchcraft* a case in which a person afflicted by witchcraft saw a vision of the suspected witch in a glass. Bernard also noted that such an apparition was "an undoubted marke of a Witch," as he had been told by a man from Cambridge who had used a glass to find stolen goods or money and claimed he could have made two hundred pounds per year—a hefty sum—with this talent.[50]

In seventeenth-century New England, Puritans were troubled by sporadic accusations of witchcraft that culminated in the 1692 Salem witchcraft trials. New England Puritan ministers Cotton Mather and his father, Increase Mather, worried and wrote about the magical and occult practices that they believed were infecting their community. Some of the practices they were concerned about involved reflective devices. Increase Mather thought witches could "shew in a Glass or in a Shew-stone persons absent." Cotton Mather railed against members of his own congregation using such devices in what they claimed were harmless practices. He believed that such practices lured people into contact with diabolical forces: "There are some that make use of wicked Charms for the finding of Secrets. . . . This is the Witchcraft of them, that with a Sieve, or a Key will go to discover how their lost Goods are disposed of. This is the Witchcraft of them that with Glasses and Basons will go to discover how they shall be Related before they Die. They are a sort of Witches that thus employ themselves." Similarly, Increase Mather attributed God's "[let]ting loose evil Angels upon New England" in part to the use of "Sieves, and Keys, and Glasses" in magical practices by people who professed to follow the Christian faith.[51]

Concerns about reflections did not focus solely on practices associated with the occult. As Puritans gained more material wealth and comfort in New England, questions also arose about the extent to which these improved material conditions would alienate the faithful from God.[52] One specific material item that Puritans warned against in this context was the looking glass. Puritan writers expressed concern about the damaging influence a looking glass might have in a woman's life by encouraging an unhealthy concern with appearance. Puritans did not invent the idea that women were too concerned with their appearance. There was already a centuries-long association in Western culture between vanity and women that had incorporated the mirror into its iconography as a material manifestation of women's preoccupation with appearance. As the French historian Sabine Melchior-Bonnet observes, vanity was one of the "metaphorical daughters of Eve," who "from the thirteenth century onward . . . is depicted brandishing a mirror."[53] But, as accurately reflective looking glasses became more widely available, Puritans worried acutely about the negative influence of this potentially vanity-inducing object.

In *Ornaments for the Daughters of Zion* (1692) Cotton Mather expressed his concern about women's preoccupation with their appearance. Mather believed women would be better served by focusing on their moral purity, devotion to God, and desire for salvation. He was particularly worried that looking glasses encouraged vanity. Mather wished that "the sex which so often looks into the Glass, would sometimes cast an eye upon this part of that Sacred *Word* which is Compar'd unto a Divine *Glass;* that they may see whether they have the Features, or Habits of, The Vertuous Woman, on them." In *Bethiah: The Glory*

Which Adorns the Daughters of God (1722) Mather warned his readers against letting extravagant dress become "a Thing of more Account" than their virtue. He admonished women to follow advice about modest dress from the New Testament, describing the scriptures as "a Looking Glass by which you do well always to dress yourselves." That looking glass would provide women with the only real measure they needed. When Cotton Mather died, in 1728, assessors recorded one looking glass in their inventory of his material possessions. Despite his protestations about the harmful properties of mirrors when used to divine the future and to consider one's appearance, even Mather found something worthwhile in possessing one.[54]

We do not know whether Lancaster's Puritan minister, Joseph Rowlandson, and his wife, Mary, owned a mirror at the time she was taken captive in 1676. Joseph died only two years later, triggering an accounting of all of the family's personal possessions. The inventory takers recorded the presence of one looking glass in Joseph and Mary Rowlandson's house in 1678.[55] Even in this rural New England Puritan minister's home, then, the looking glass had made its appearance. It seems likely that this was an item the Rowlandsons might have first acquired postcaptivity, when their status and wealth increased substantially. It is too tempting not to speculate that using Quinnapin's mirror inspired Rowlandson to acquire one of her own. In the rural area of Connecticut where they lived in 1678, in any case, owning a mirror set them apart from the vast majority of the residents of Wethersfield, where mirror ownership in the 1670s reached only 17 percent of households.[56]

Like Quinnapin, Rowlandson came from a tradition that had much to say about reflective power and the uses for reflective surfaces. Divination had been widely practiced in England and had survived the Atlantic crossing. Reflective objects were a widely recognized tool, not only of witches in New England and those who sought to denounce them but also of ordinary Puritan believers who sought to know their futures. Rowlandson also worshipped in a faith that encouraged her to downplay her attention to appearance and to eschew the looking glass because it promoted vanity.

On that day in 1676, when Quinnapin offered Mary Rowlandson his mirror, it was almost certainly true that mirrors were a more commonly owned item in local Native society than among the English in this rural area of Massachusetts. As historian James Axtell has broadly described the "first consumer revolution" among seventeenth-century Native Americans, the "Indians of the Eastern Woodlands experienced a consumer revolution . . . many years earlier" than Europeans did, "usually as soon as the commercial colonists founded trading posts, *comptoirs*, and nascent settlements."[57] In just one decade of trade, between 1652 and 1661, Pynchon had enough mirrors available for ownership rates among individual Native peoples to reach 31 percent (and much higher if mirrors are assumed to have been shared by several members of the same family) while ownership rates among white families in nearby Wethersfield, Connecticut, reached only 17 percent of households in the 1670s. That Mary Rowlandson showed no surprise that Quinnapin had a mirror suggests that she knew of their popularity as a trade item or had seen them in use by Native peoples during her captivity. Despite Puritan misgivings about women's use of looking glasses and their concern about the connection between glasses and witchcraft, Rowlandson hesitated not a moment when Quinnapin offered his to her. She raised no questions about its intended use, understanding that Quinnapin gave it to her so that she could see clearly the poor state of her appearance and remedy it.

The mirror used in this exchange between Quinnapin, a Narragansett man, and Mary Rowlandson, an English Puritan woman, was an object freighted with multiple meanings in both Narragansett and Puritan cultures. As an object of potential power it commanded respect; as an instrument of vanity it suffered scorn; as a useful item of the material world it was embraced by men and women in both Native and English cultures. None of these layers of meaning would disappear from peoples' interactions with mirrors as the next two centuries of life unfolded in North America. Moreover, the idea that the mirror was a vital source of information about the self—Quinnapin "gave me the glass to see how I looked," as Rowlandson affirmed—would make this object a powerful tool and symbol of personal and collective identity in North America.

CHAPTER THREE

Looking-Glass Ownership in Early America

In 1832 the American Providence Exploring Company sent the brig *Agenoria* and its seventeen-man crew to explore the River Niger in Africa and trade for a wide variety of goods.[1] The *Agenoria* carried thousands of items to exchange for these desired goods, including iron, cloth, soap, rum, tobacco, umbrellas, knives, mirrors, boxes, dishes, scissors, and jewelry. The *Agenoria*'s mirror supply consisted of fifteen hundred "Pocket Looking Glasses," valued at $0.35 a dozen, and 144 "Toilet Glasses," valued at $1.12 a dozen. In addition to a meticulous list of trade goods that the *Agenoria* carried, the ship's secretary kept a careful record of items purchased by the crew. From that accounting, we know that four of the seventeen men purchased looking glasses for themselves during the expedition. When the *Agenoria* returned from its voyage, a portion of its cargo remained unsold, including some mirrors. Given the rigorous record-keeping heretofore, it was somewhat unexpected that no one recorded how many mirrors and some of the other items remained at the voyage's end. It had, however, been a very difficult journey. Only seven of the crew of seventeen had survived.[2]

At first glance, the records of the *Agenoria*'s voyage provide us with a surprising amount of information about mirrors aboard a nineteenth-century American merchant ship. We know that at least 24 percent of the ship's crew owned a mirror because they purchased one during the voyage. The 1,644 mirrors included among the trade goods suggest that this item was in demand among West African consumers, although not so in demand that they sold out during the voyage. Not only were mirrors well-supplied on the *Agenoria* compared to the quantities of other trade goods aboard, but by the 1830s—centuries after Europeans first brought material goods to Western Africa—it was unlikely that an item would have been included if there was not an established market for it. Although these mirrors were not destined for consumers in America, the *Agenoria* traded its goods among African populations from whom millions of men, women, and children were captured and forced into slavery, so an interest in this item suggests potential familiarity with and a similar interest in mirrors among people from these regions enslaved in North America.

The types of documentation kept by the *Agenoria*'s crew—trading records and inventories of personal possessions—are the two kinds of evidence most used in this study to track ownership of mirrors among peoples of African, Native, and European descent in North America. Trade records and information about individual ownership of items provide the most robust bodies of evidence for the historian seeking to understand patterns of ownership of material goods. But trade records, like those of the *Agenoria*, often leave important questions unanswered. Commonly missing, as in this case, is information about to whom the mirrors were sold and how many remained unsold. Records of individual ownership can similarly disappoint. The *Agenoria* recorded only what the men purchased while employed by the ship, not what they might have brought with them when they boarded the vessel. This problem is largely overcome in probate inventories—the most common records detailing individual white ownership of material goods in early North America—which were taken shortly after an individual's death and were meant to list the entirety of a person's worldly belongings in order to assess the value of the deceased's estate. But even though probate records were intended to be comprehensive, problems arose. In some locales, for example, it was traditional to exclude "family items that were not truly the possessions of the [male] decedent, such as a wife's personal effects," which led to the exclusion of some belongings from the records.[3]

An even more fundamental issue arises in using trade records or inventories of personal possessions to track interaction with mirrors in early America: mirrors did not have to be owned to be used. We have only to think here of Mary Rowlandson's use of Quinnapin's mirror as an example. One mirror could also be used by many members of a family or by many crewmen on a ship. And the homes of wealthy Americans, which showcased multiple and ostentatious accurately reflective glass mirrors, were occupied not only by their owners but by men, women, and children in service or enslaved to the owners. Servants or enslaved persons working in one of the home's public spaces, or standing behind a wealthy master in a private chamber looking into a mirror to ensure that the master's clothing or hair was styled correctly, also saw their own reflections in the looking glass. Moreover, because mirrors did not have to be touched to be used, this was an item that anyone could employ without worrying about breakage or accusations of theft, which could have serious, even fatal, consequences for a slave. How many early Americans had access to looking glasses they did not own cannot, of course, be known. But it is important to note that mirrors, as visual objects—items that could be used without having to be touched—could be used more widely than just by their owners.

As the example of the *Agenoria* and the preceding discussion illustrate, tracing patterns of mirror ownership among early Americans presents many challenges. Nevertheless, a necessary first step in understanding any group's interest in, use for, and significance given to a material object is attempting to understand that object's penetration and circulation within the group. This chapter seeks to quantify mirror ownership among early Americans wherever that is possible but also identifies avenues for acquisition and interest in mirrors, especially when quantification is difficult.

Africans and African Americans

Attempting to understand interest in mirrors and the incorporation of them into the lives of people of African descent calls us to consider evidence on two continents. European explorers, slavers, and traders gained access to West and West Central African nations during the centuries-long era of the transatlantic slave trade, which enslaved an estimated 12,521,000 Africans between 1501 and 1867.[4] These Europeans often recorded their thoughts and observations about the places they visited and the people they encountered. Seeking evidence of interest in mirrors, meanings associated with reflection, and uses for reflective materials in these European encounters with people in Africa offers the possibility of insight into the worldviews of cultures whose members would be victims of the Atlantic slave trade. This evidence is especially important because we have very little material culture evidence and no probate or trade records that document patterns of ownership of mirrors among Africans and people of African descent in North America. Instead, we must rely on archaeological evidence and more anecdotal material that reveal avenues of acquisition and access.

By 1600, many Europeans regularly included mirrors in their cargoes as potential gifts or trade items for the nations of coastal West Africa, which was the homeland of 50.2 percent of the captive Africans in the transatlantic slave trade and was the homeland of 75 percent of those captives who left Africa on ships for North America.[5] Descriptions of Ghana (Guinea) by Dutch trader Pieter de Marees (1602) and by Germans Andreas Joshua Ulsheimer (1603–4) and Wilhelm Johann Müller (1662–69) all noted that mirrors were popular trade items.[6] London export records show 216 "coarse" glass mirrors (likely the earlier, less accurately reflective ones) destined for Guinea in 1634 and 420 small "crystal" glass mirrors (the more accurately reflective kind) headed to Guinea and Barbary in 1640.[7] The Frenchman Jean Barbot's observations in Guinea at the beginning of the eighteenth century reveal that mirrors were available by then in different varieties, including "gilt or plain wooden frames." Olfert Dapper's description of Benin confirms that mirrors were being traded there as well by the mid-seventeenth century.[8] When the Englishman John Atkins made his first slaving voyage to West Africa in the early eighteenth century, he found a strong demand for mirrors. The merchandise he brought with him from London fell short, as he acknowledged, owing to his lack of experience: "I was but a young Trader, and could not find out till I came upon the Coast" that he had brought items of little interest to those at the factories who held captive Africans for sale. His original, inadequate inventory contained no mirrors, but he included looking glasses on the list, learned from experience, of "what Goods are asked for."[9] The observation of one English slave trader, John Matthews, made around 1800, does suggest that mirrors had at least become a common item among high-status women in the West African coastal nation of Sierra Leone, writing that while "common people" slept outside around a fire, "people of consequence have bed places, made by driving four stakes into the ground, with a bottom of split cane or bamboo; and mats hung round supplies the place of curtains." The women stored "their domestic utensils, mats and stools" inside this housing, which was, according to Matthews "never without a looking-glass."[10]

The area of West Central Africa engaged in the slave trade stretched southward from "Cap Lopez (Gabon), at the southern end of the Gulf of Guinea to the southern tip of Africa." West Central Africa was the homeland of 45.5 percent of the captive Africans in the transatlantic slave trade, and was the homeland of 23 percent of those captives who left Africa on ships destined for North America.[11] Less is known about the trade in mirrors in this region, although according to one study, mirrors were a popular item in West Central Africa as early as the seventeenth century.[12]

Several nineteenth-century European explorers' accounts in the interior regions of West Africa provide evidence, if inconclusive, of interest in mirrors and individual ownership of these objects. In 1823 the English explorer Dixon Denham (1786–1828) undertook an expedition that began at Tripoli, on the Mediterranean Sea, and would eventually reach Lake Chad in the interior Kingdom of Bornou. Denham observed a vigorous demand for "English goods," such that even the cheapest quality items were eagerly sought after and commanded exorbitant prices. He included "small looking-glasses" among the "articles most in request" by African nations. After Denham distributed all of the mirrors he had brought with him for the journey, he continued to encounter people who wanted this item. One man, a Sultan, asked for a looking glass from Denham, who was frustrated that he did not have one to offer, especially since he knew they had "cases" of the desired goods among the supplies they had left behind at Bornou. Pondering the request, Denham acknowledged to himself that the only looking glass he had remaining was his own. But not even the sultan's high status nor Denham's desire to satisfy the request could convince him to part with his own small mirror.[13]

When they did have a supply of mirrors on hand, European explorers often included looking glasses in gifts they gave to high-status individuals and people who provided them assistance on their journeys. The Scotsman Hugh Clapperton, with whom Denham began his expedition, until the two men parted ways near Lake Chad at Kuka, noted giving looking glasses to three men who had led a party dispatched to find the expedition's lost camels. In thanks for the safe return of the animals, each man received "a cotton kaftan, or loose gown, a knife, looking-glass, snuff-box, razor, and some spices."[14]

Experiences recounted by the expedition's leaders suggest that exposure to and ownership of mirrors varied by locale. Denham described an encounter with a group of women who were not familiar with their mirror selves. After separating from Clapperton, Denham's expedition traveled to a town called Yeddie, along the southwest side of Lake Chad. A group of curious onlookers soon gathered outside the expedition's lodging. Denham implored the local kaid to send them away so that the travelers might rest. During his conversation with the kaid, Denham mentioned that he had noticed that all of the local inhabitants who had gathered around the hut were men, so he asked whether there were any "women in your town?" Denham was told that, of course, there were; his afternoon was spent, as a result of this question, receiving more than one hundred local women who came to see him. Denham seemed contrite that he had little to show them except a mirror, which he put to good use to pass the time. Denham asserted that "probably nothing could have pleased them more." He observed two women who brought other female relatives with them so that each could see "the face she loved best reflected by the side of her own, which appeared to give them exquisite pleasure; as on seeing the reflection they repeatedly kissed the object of their affection." Similarly, a young mother brought her child to the mirror with her. The mother, according to Denham, "absolutely screamed with joy; and the tears ran down her cheeks when she saw the child's face in the glass, who shook its hand in token of pleasure on perceiving its own reflected image."[15] In contrast, however, Clapperton observed, both in his journey with Denham in Bornou and his own travels into the Sokoto Caliphate (present-day Nigeria), how he had seen women who "dye their hair blue, as well as their hands, feet, legs, and eyebrows," using a handheld looking glass to assist with the paint's application. Although one mirror could be shared among several women, Clapperton seemed to suggest here a wider distribution of mirrors.[16]

Clapperton and Denham reunited at Kuka and made their way home to the United Kingdom in 1825. Scarcely three months later, Clapperton returned to Africa, this time landing along the coast of West Africa at Badagry in the Bight of Benin. From there his expedition traveled north through the Yoruban homeland (present-day Nigeria) to Sokoto. Clapperton intended to go farther west and return to Bornou, where he had been with Denham, but he died at Sokoto. On this final voyage to Africa, Clapperton visited the home of a wealthy woman, the widow Zuma, and noted that she owned two looking glasses, which she had brought before him with other possessions to showcase her wealth and status. The expedition also gave gifts that included mirrors on two occasions—to a king and a sultan. The records from this voyage also provide a very rare glimpse of what one of these mirrors looked like. This mirror was owned by a woman who used it frequently, according to Clapperton. He described the mirror as "a circular piece of metal, about an inch and [a] half in diameter, set in a small skin box." Clapperton seems to have described almost perfectly one of the tin mirrors made in Europe during the seventeenth and early eighteenth centuries for trade with Native Americans. The leather box could have been fabricated in Europe or Africa. Whether these mirrors were widespread in Africa remains unknown.[17]

Richard Lander was the only European to survive Clapperton's expedition, but this did not deter him from returning at the end of March 1830 with his brother John to explore the Niger River. The Lander expedition brought material goods to distribute on their journey, including fifty razors, fifty pairs of scissors, sixty knives, sixty-four silver armbands, one hundred combs, one hundred common snuff boxes, and 110 mirrors (one hundred of which were noted for an unknown reason as being of "inferior quality"). The only thing they brought in great numbers were sixty thousand assorted needles, with which they found their trading partners already oversupplied. Lander noted several occasions on which he gave mirrors as part of gift packages to high-status Africans, including a "governor and his minister," a "king," and a "king and queen." When the gifts were given to this king and queen, at Boossa along the River Niger, the king was particularly interested in "the pair of silver bracelets, a tobacco-pipe, and a looking-glass," which "seemed to rivet [his] attention, who could not take his eyes off them for a full half-hour, so much was he pleased with them." By the middle of October, the expedition had run through almost all of their goods, having only "needles and a few silver bracelets left to present to the chiefs." Lander did observe a practice common in many parts of the world but uncommon in early America: using small shards of mirrored glass to adorn clothing and other textiles. The Lander expedition met the "King of the Eboe country," who wore "a cap shaped like a sugar-loaf," and "covered thickly with strips of coral and pieces of broken looking-glass."[18] Although Lander commonly noted giving gifts to high-status people that he met on his journey, he made no mention of participating in such a ritual with the King of the Eboe country.

Two years later, Scottish merchant MacGregor Laird, accompanied by Lander, undertook another journey up the Niger River into the Igbo region (what Lander called the "Eboe country"), similarly giving high-status men and women he met along the way looking glasses as gifts.[19] Like Lander had on his previous voyage, Laird met with the King of the Eboe country. Upon first encountering him, Laird observed one present that the King had with him that had been given to him by Lander—so it is likely that Lander did provide the King with some gifts, although they went unrecorded. Laird's gifts for this king included an arm chair "lined with scarlet cloth," a large looking glass, two jars of rum, and some cloth and clothing. Upon receiving the gifts, the King "examined [the chair] very attentively, and then seating himself in it, called for a looking-glass. He then proceeded to examine himself in it, and burst out into a loud laugh, thinking himself, no doubt, the happiest of monarchs. He remained thus surveying and laughing at himself alternately for some time." Even though Laird had brought him a large looking glass, Laird noted that once he received the chair, the King "called for a looking-glass," presumably one he already owned (because Laird does not identify it as the one he had brought as a gift). It was likely the chair, and not the looking glass, that gave this king such great delight. But the looking glass was critical for the King to be able to observe himself seated in the chair and to incorporate how he looked there into his understanding of himself.[20]

Looking glasses were popular more widely among the Igbo people. Laird specifically observed that the Igbo "manifest a great partiality for rum, small looking-glasses, and cowries." Material culture evidence from the Igbo region suggests that these small looking glasses were desired so they could be put into native-made wooden frames. The exhibit *Igbo Arts: Community and Cosmos*, at the Museum of Cultural History at the University of California, Los Angeles, gathered at least eight examples of intricately carved small wooden mirror frames that "were often carried by titled women as display items in rites of installation, but they were not restricted to them." These frames vary in shape, but all have a handle and, on all but one, on the end of the handle is a curved, triangular, or rectangular open piece of wood into which the mirror owner's hand could fit to hold the looking glass.[21]

Additional carved wooden mirror frames from Nigeria are held in the collection of the Pitt Rivers Museum at Oxford University, England.[22] Three rectangular, flat-glass mirrors were housed in wooden frames. Unlike the Igbo mirrors, these all have sliding wooden covers to protect the fragile mirror glass. One of these had been collected in 1909 from Nigeria (fig. 3.1) and the other two in 1925, one from Nigeria and one from Ghana. Another, from Nigeria, is a wooden fan, housing a small, slightly irregular rectangular piece of flat mirror glass, collected in 1884 (fig. 3.2). The Pitt Rivers collection also includes a 1.5 in. diameter circular glass mirror in a hide casing, collected in 1884 (fig. 3.3), that in shape and size appears to be similar to the metal mirror described by Clapperton.[23]

The extent of mirror acquisition and use among people of African descent in North America resists quantification, although avenues of access and acquisition can be demonstrated. Although ownership of mirrors among African Americans cannot be quantified, the aggregation of archaeological evidence is beginning to provide some critical data. The Digital Archaeological Archive of Comparative Slavery includes 297 pieces of mirror glass recovered at eighteenth and nineteenth century plantation sites in Maryland, North Carolina, South Carolina, Tennessee, and Virginia. These pieces of mirror glass were recovered from structures associated with African Americans at 52 percent of the plantation sites in this archaeological archive.[24] Not included in this count are the mirrors found at Portici Plantation, built in 1820 in the piedmont region of Virginia, which was destroyed by fire during the Civil War. The nineteenth-century slave quarters at Portici, known in project reporting as Structure 1, had three fragments of silvered glass, which the archaeologists

Figure 3.1. Mirror frame with sliding cover (Edo; Nigeria, Benin), no. 1909.67.14.1 and 1909.67.14.2. Copyright Pitt Rivers Museum, University of Oxford. Used by permission.

noted were "probably from a small hand-mirror." The remains of the plantation house cellar, where enslaved workers lived, contained forty-four fragments of mirror glass, and the findings from the material culture associated with the planter family's portion of the house included 974 fragments of mirror glass. Fragments of mirrors were also recovered at the Levi Jordan Plantation site in Brazoria, Texas, which was established in 1848.[25]

These archaeological remnants document mirrors in African American dwellings but remain silent on how mirrors were acquired or accessed. Enslaved men and women had three main avenues for the acquisition of goods: mirrors might be supplied by a master, acquired through the internal economy, or gained by illicit appropriation. Moreover, as John Michael Vlach has observed, enslaved men and women acquired goods over time and passed them down: "since generations of slaves were raised in the quarters, there was necessarily an accumulation of ragged and meager artifacts." Free people of African

Figure 3.2. Mirror fan (Nigeria), no. 1884.70.23. Copyright Pitt Rivers Museum, University of Oxford. Used by permission.

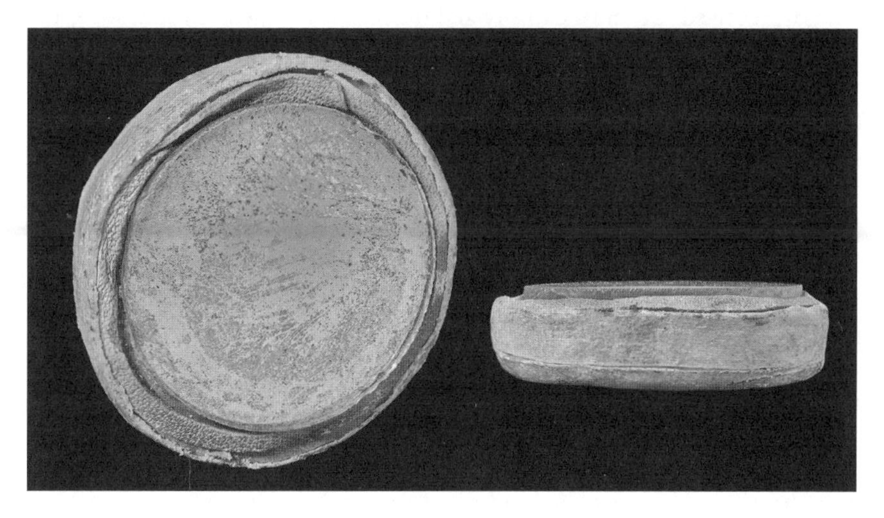

Figure 3.3. Small circular mirror set in a hide frame (Africa), no. 1884.70.20. Copyright Pitt Rivers Museum, University of Oxford. Used by permission.

descent could similarly acquire mirrors as gifts, by purchase, through theft, or by the generational transfer of goods.[26]

Masters gave their slaves looking glasses for a variety of reasons. Masters might show material beneficence in an attempt to produce goodwill among an enslaved workforce. One antebellum North Carolina slave owner supplied "a cheap looking glass" for each slave dwelling on his plantation. He hoped it would increase their feelings of goodwill toward him. His intention of "improving the slave and uniting him to his owner" apparently worked, at least from the master's perspective, who noted with satisfaction "a marked change for the better" among his slaves. Some masters provided their enslaved workforce with looking glasses in order to ensure a certain level of acceptable appearance on the part of the enslaved. Providing the mirror put the burden of regulating the master's standards of acceptable appearance on the slaves rather than the master. Despite the Spartan surroundings of the slave dwelling in which he had spent his childhood, George Fleming recalled, for example, a looking glass that had been given to his family by "Missus Harriet," "so we could see how to fix up." Masters could also bequeath material goods at the time of their death. When Eve Scurlock of New York City died in 1754, she freed her five slaves. Along with his freedom, one of the enslaved men, Caesar, received "4£ and a pair of hand irons, and 1/2 the firewood, soap, and candles, six plates, the English books, and a small looking glass."[27]

Many of these looking glasses may have been ones that masters proffered because they had broken and outlived their usefulness in the owner's house.[28] As Levi Pollard's detailed description of the house he had lived in while enslaved suggested, slaves may have commonly lived among items that had been deemed no longer suitable for their masters' houses, what Pollard called the "white folks throwoffs." Slave dwellings sometimes had a "piece" of a looking glass, suggestive evidence that masters may have supplied broken pieces of looking glasses from their houses to slave dwellings, thus avoiding any additional expense while still providing this item. The British actress Frances "Fanny" Kemble, visiting Georgia in the 1830s, for example, described one "broken looking-glass" in a slave dwelling. The "old piece of broken looking-glass" and the "dresser wid a piece of looking glass to look in" that Jacob Green and Anne Maddox recalled from their dwellings under slavery may well have been these kinds of "throwoffs" from their masters' houses as well.[29]

People of African descent also bought mirrors. Enslaved men and women could be involved in the internal economy or gain access to money through some other effort. They could become active purchasers of material goods, joining their free counterparts at a range of venues.[30] Account books reveal that enslaved men and women purchased looking glasses at local stores. Historian Ann Smart Martin found that among the enslaved population who frequented John Hook's store in late eighteenth-century backcountry Virginia, "inexpensive mirrors were popular." Estate sales also offered opportunities to acquire consumer goods. Sam Williams, an enslaved worker at William Weaver's Buffalo Forge in the Valley of Virginia, purchased a looking glass at such a sale in 1851, paying $1.75 for it. Newly freed African Americans on St. Helena Island, South Carolina, who were part of the Port Royal Experiment during the Civil War, showed a strong interest in purchasing consumer goods. One item available for purchase was "a quantity of very small, low-priced looking glasses" that came "into immediate request" and, apparently, had all been sold.[31]

Looking glasses could also be acquired through theft, but this seems to have been uncommon. John Joyce, a free African American man living in Pennsylvania, who confessed to theft and murder before being executed in 1808, admitted to having stolen several items from his victim's house, including "a bundle of clothes, and a looking glass." A study of antebellum upcountry South Carolina's Court of Magistrates and Freeholders, however, found that out of 234 cases involving property theft by slaves, only one included a stolen looking glass. While enslaved men and women used illicit appropriation to acquire material goods, looking glasses may have been an unappealing object of these kinds of activities. In addition to the possibility of breakage, any but the smallest looking glass would have been hard to conceal and, in many households, quickly missed by its owners, especially if it had been prominently displayed.[32]

Enslaved men and women could also make use of the mirrors in their master's house, if they had access to them. Some recalled fairly frequent access to white-owned mirrors and the opportunity to come to know their mirror selves in their masters' houses. In their Works Progress Administration (WPA) interviews recorded in the 1930s, sisters Pauline Johnson and Felice Boudreaux recalled going "up to the missy's looking-glass to see if us pretty enough go to church" every Sunday. Daphne Williams's recollection suggests a familiarity with the image she saw reflected in her mistress's mirror: "The missus was a widow woman ever since I 'member her. She have two boy and three gal, and that sho' was a lovely house. They have they ownself painted in pictures on the wall, jus' as big as they is. They have them in big framce [*sic*] like gold. And they have big mirrors from the floor to the ceilin'. You could see you ownself walk in them."[33]

Avenues of acquisition and access provide critical information about the ways in which people of African descent in early North America could have become familiar with their mirror selves. Some African Americans later recalled a time in their lives before they had ever encountered their mirror selves. In her WPA interview, Millie Manuel detailed the spartan living conditions she had endured under slavery: "We had timber rail house. No beds in it. We slep' on the floor on a pallet. We didn't have no chair and we didn't have no mirror." That last item—the mirror—seemed particularly important to Manuel, who continued, "I didn't knowd what I looked like 'til I was free." Alexander Robertson similarly talked about how, after freedom, he had the opportunity to "'pear in de lookin' glass." Linking seeing one's mirror self for the first time with freedom suggests the limited access to mirrors some enslaved men and women had. It also presented the idea that one could not own one's mirror self—as one could not own one's body—until freedom had been achieved.[34]

Native Americans

The most comprehensive records that enable us to apprehend Native peoples' ownership of looking glasses in North America come from the British fur trade in eighteenth-century Canada. While the primary focus of this study is the region of North America that would become the United States, Native cultural groups occupied both sides of what became the border between Canada and the United States. The border thus does little to help us recognize the common identities it bisected.[35] Moreover, the unique nature of the records available from the British trading presence in Canada make this an invaluable source for the present study.

In 1670, as British fur-trading interests shifted farther north, where the supply of animal pelts remained robust, England established the Hudson's Bay Company. The Company built trading posts around the southern and western shores of Hudson Bay and James Bay in the Canadian subarctic to which Native trappers brought furs to trade. The Canadian subarctic was inhospitable to agriculture and sparsely populated. The Native peoples who traded with the Hudson's Bay Company included the Cree, Assiniboine, and Chipewyans who lived in vast hinterlands that extended out beyond the trading posts.[36] The trade at Hudson's Bay Company posts was based on the "made beaver" standard (MB), which represented the value of one "prime beaver skin." All products (and other furs) traded at the posts were priced according to the number of MB needed to acquire them.[37] Hudson's Bay Company employees kept voluminous and meticulous records that provide uniquely detailed information about many aspects of the trade, including not only the number of goods the company stocked but also how many of each item Native consumers purchased every year.

Mirrors were a part of the Hudson's Bay Company inventory from the outset. In 1674, the company's inaugural trading year, an order was placed for 864 looking glasses—576 "tin Lookeing glasses" and 288 "Small lookeing glasses"; an additional 1,728 "small painted glasses" followed in 1682.[38] Although these were the Hudson's Bay Company's initial forays into trade with Native peoples, the company relied on knowledgeable European traders with extensive experience in North America to determine what goods they should stock.[39] As company employees spread out into the Canadian subarctic to establish relationships with Native communities during these early years, they carried with them sets of goods they stocked to create interest in their products, including looking glasses.[40]

In the early years of its existence, the Hudson's Bay Company stocked metal and glass mirrors. Leather looking glasses (what Pynchon called "book looking glasses" in New England) were popular here, although the company stocked other kinds of looking glasses, including a square variety. There is also evidence that the company stocked metal mirrors. In the late seventeenth-century and early eighteenth-century ledgers of the Hudson's Bay Company, an item that almost always appeared immediately before or after the company's supply of glass mirrors was something called a "tinn showe" or "tin show" (fig. 3.4). I believe that Hudson's Bay Company employees called metal tin-plate mirrors tin shows, as in tin that "shows" a reflection.[41] Its evocative name and placement in the company records strongly suggest that this item was a metal mirror. Moreover, as the supply of tin shows petered out at Hudson's Bay Company posts in the early eighteenth century, the number of looking glasses sold

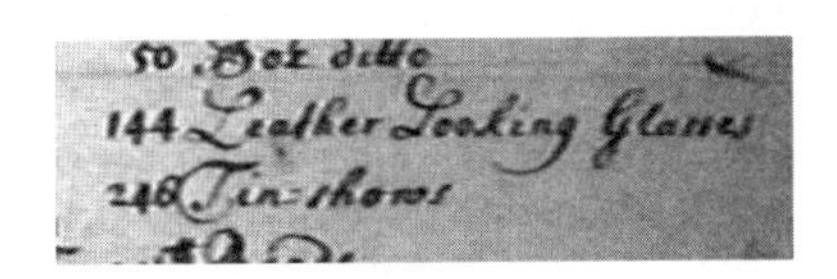
144 Leather Looking Glasses
248 Tin shows

Figure 3.4. Entries for "Leather [Book] Looking Glasses" and "Tin shows" in the Fort Albany Account Books, 1698–99, B.3/d/10. Microfilm Records no. 1M407, Hudson's Bay Company Archives, Archives of Manitoba.

increased, suggesting that consumers who previously would have purchased metal mirrors were now buying ones made of glass.

At the Hudson's Bay Trading Post of Fort Albany, in 1708 and 1709 there were, for example, 144 "book looking glasses" in stock, 101 of which were sold. That season, company employees at Albany also stocked 216 tin shows and sold 120 of them. In 1709 and 1710 no new mirrors arrived; of the remaining mirrors from the previous year, the Fort sold four book looking glasses and forty-eight tin shows. During 1710 and 1711 no new mirrors arrived again, but the Fort sold all of its remaining supply: thirty-nine leather looking glasses and forty-eight tin shows. The following year no mirrors of either variety arrived, so no mirrors were available for sale. In 1712, only glass mirrors were included in the shipment: eighty-four "book looking glasses" and a new variety, thirty-six "square looking glasses." No tin shows arrived in 1712 or following years. At Fort Albany, after the supply of tin shows was exhausted, the average number of glass mirrors sold annually rose from forty-four to sixty; some of the people who would have bought tin shows now seem to have bought looking glasses.[42] As looking glasses replaced metal mirrors across Europe and North America by the early eighteenth century, these small metal mirrors disappeared almost entirely from the memory of this earliest era of the fur trade in North America, although they were an important item in their time.

The direct evidence about the sale of mirrors to Native consumers served by the Hudson's Bay Company comes from individual company trading posts. The two most significant of these trading posts were Fort Albany and York Factory.[43] Fort Albany, located on the western shore of James Bay (a smaller bay that descended from the southeast corner of Hudson Bay), had a hinterland that included "the lands lying just to the east of Lake Winnipeg" and an estimated population of 3,450. Between 1698 and 1717, Fort Albany stocked 1,494 mirrors and sold all of them. In the mid-eighteenth century, York Factory, located on the southwest corner of Hudson Bay, had an estimated eighty-six hundred people living in its much larger, almost four-hundred-thousand-square-mile, hinterland, which extended into "much of present-day Manitoba and part of Saskatchewan," as well as "a portion of northwestern Ontario." At York, between 1716 and 1770 and between 1775 and 1782, the English traded 6,271 mirrors.[44]

The combined information about population and numbers of mirrors at Fort Albany and York Factory allows us to estimate mirror distribution among these populations. In the earliest years of trade with the Hudson's Bay Company, one mirror was sold for every 2.31 people in the areas served by Fort Albany. As the trade matured throughout the eighteenth century, York Factory supplied its larger consumer base with enough mirrors to average one for every 1.37 people in its trading regions. If mirrors are taken to have been available to multiple members of the same family, then rates of mirror ownership would be significantly higher if counted by family unit.

Two mitigating factors deserve mention, both of which impact this estimate of how many mirrors were in these Native communities served by Hudson's Bay Company in the eighteenth century.[45] On the one hand, historian Arthur Ray has shown that the Assiniboine and Western Cree, who lived in the York Factory hinterlands, transferred a "large proportion of the trade goods" they acquired "to other groups with whom the two tribes traded," meaning that some mirrors undoubtedly passed through the York Factory hinterlands on their way to other destinations, reducing the total number of mirrors in circulation in the areas served by the Hudson's Bay Company.[46] On the other hand, English-traded mirrors were not the only ones entering the region. From the outset the Hudson's Bay Company faced significant competition from the French. The French developed a different model for trade than had the English. Instead of establishing trading posts to which Native trappers brought their furs, the French sent their traders out into the hinterlands to trade with Native peoples in their own communities. The French, by necessity, had to focus on material goods that the traders could transport across the vast hinterlands. Thus the French competed with English traders largely over the lighter, smaller items that could be more easily transported, including mirrors. Early in the century, the location of York Factory (west of Hudson's Bay) largely protected it from competition with the French, who were based far to the south and east at Quebec City and Montreal, but beginning in 1730, French competition emerged in the York Factory hinterlands and increased through the 1750s. The French accounted for 40 percent of the trade in the York Factory Hinterlands by the 1760s. At Fort Albany, French competition was much fiercer because of its more easterly location compared to York. At York Factory the standard of trade was one MB traded for one mirror, but at Fort Albany, where prices had to be more competitive, one MB bought two mirrors. Although we do not know how many mirrors the French introduced into the Canadian subarctic during the eighteenth century, all available evidence suggests that it would have been a substantial number.[47]

The trading years of 1775 to 1782 were appended to the end of the data collected for York Factory because they enable us to assess one additional component of the trade in this area—namely, how evenly distributed mirrors were across the vast hinterland served by this trading post. In 1774 Hudson's Bay Company opened a new trading post, Cumberland House, in the York Factory hinterlands, west of Lake Winnipeg. Cumberland House quickly came to serve an estimated 70 to 80 percent of the Native traders who had previously traveled to York Factory. This left only the 20 to 30 percent of the traders living close to York Factory who continued to bring their furs there to trade. In the 1970s, historian Arthur Ray undertook a study of guns and powder, kettles, hatchets, knives, broadcloth, blankets, beads, tobacco, rum, and brandy at York Factory. One of the questions he explored was whether the abrupt change brought on by the opening of Cumberland House in 1774 revealed any different patterns in preferences for goods between those Native peoples who lived near York Factory and those who were dispersed into its hinterlands. He found that those living closest to York Factory "were buying large quantities of blankets, cloth, and beads," while further from York Factory, "chiefly the Parkland Assiniboine and Western Cree were the principal consumers of tobacco," while kettles were fairly evenly distributed across all populations who traded at York Factory.[48]

In the 1760s, York factory traded an average of ninety-seven mirrors each year. Between 1775 and 1782, York traded an average of twenty-three mirrors per year—24 percent of the 1760s average—closely reflecting the shift of population away from York Factory and showing that the mirror was an item that people near the fort and in its hinterlands consumed equally and that it was therefore probably well distributed throughout the populations living in this vast hinterland. This even distribution of mirrors also suggests that the Assiniboine and Western Cree were not primarily buying up large quantities of mirrors in order to trade with other groups outside the reach of the Hudson's Bay Company.[49]

While the records for the eighteenth-century Canadian subarctic, combined with those of seventeenth-century New England (discussed in chapter 2), allow for a quantitative assessment of looking-glass penetration into these Native societies, the more impressionistic and anecdotal evidence available from across North America suggests that New England and the Canadian subarctic were representative of broader patterns, not outliers. As Europeans expanded their settlements in the eighteenth century, they continued to include mirrors among the items they traded with Native peoples. In 1701,

French suppliers sent 432 "medium-sized looking glasses" for trade in Louisiana. A 1736 South Carolina store inventory included sixty looking glasses marked for "Ye Indian Trade." Traders in the Great Lakes region stocked looking glasses at midcentury.[50] Native trade networks that reached far beyond the outer edges of European settlement undoubtedly carried mirrors deeper into the interior than European records can reveal.

Sometime around the middle of the eighteenth century, a new variety of mirror began to appear in the trade records. This mirror was called "paper framed," "pasteboard frame," or simply "paper." These small, circular looking glasses had a thick paper backing (a forerunner of cardboard). The glass and paper were affixed to one another by a thin metal frame. Easily perforated by a needle and thread, the paper provided a surface that allowed these mirrors to be attached to clothing or other items. Whether the idea to create this novel form originated with demand from Native consumers or European suppliers is unknown, but in either case, paper looking glasses quickly became a popular item among Native consumers, and Native demand undoubtedly drove European production of this item. The precise timing of the introduction of this item remains obscure. The earliest possible evidence appears in the record of a Montreal trader from 1738, who had mirrors with "red paper frames" in stock.[51] In any case, a robust trade in paper looking glasses had emerged by the late eighteenth century in North America. In 1778 the "List of Goods on Hand for the Indian Department" of the United States at Detroit included twelve hundred "paper Looking glasses."[52] At York Factory, Hudson's Bay Company stocked 296 "gilt paper looking glasses" between 1775 and 1782. In the early nineteenth century the American Fur Company stocked a large supply of paper looking glasses. John Jacob Astor had founded the American Fur Company in 1808 in New York, but by the 1820s it had expanded to the Great Lakes, St. Louis, and the Rocky Mountains and had made Astor "the single most powerful fur trader in the United States." In the eight-year span between 1828 and 1836 the American Fur Company purchased 7,560 paper-covered looking glasses.[53]

In the Plains region during the nineteenth century, traders and settlers continued to mention the importance of mirrors for trade. Pierre Tabeau's description of his and French Canadian trader Regis Loisel's 1803 expedition into the region of the Upper Missouri River included a discussion of the trading goods necessary for engaging successfully with the Arikara. Tabeau noted that "ammunition, knives, spears, blue beads, tomahawks, and framed mirrors are the only articles for which they are willing to exchange their robes." The emphasis here on "framed" mirrors raises the question of whether what the Arikara were rejecting (which, presumably, other Native groups were not) were either mirrors entirely unframed or those paper looking glasses, which had the least substantial mounting. At midcentury, settler Margaret Frink described an encounter with a group of Sioux near Scott's Bluff, Nebraska, during which "mirrors in gilt frames, and a number of other trinkets" were exchanged for

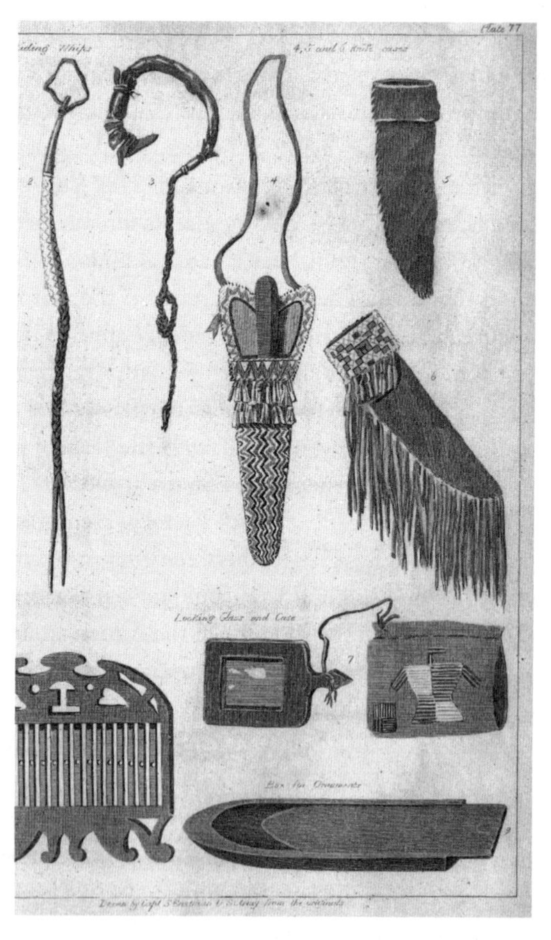

Figure 3.5. Looking glass and case (*bottom right*). Drawn by Capt. S. Eastman U.S. Army from the originals. Henry R. Schoolcraft, *Information Respecting the History Condition and Prospects of the Indian Tribes of the United States . . .* (Philadelphia: Lippincott, Grambo, 1853), plate 77, *Implements* (detail).

"fish and fresh buffalo, deer, and antelope meat." Although Frink had money with which to buy this food, the Sioux were not interested in it. In fact, they "would not look at" the money; the goods were necessary for the exchange.[54]

Native peoples not only acquired European looking glasses; they also modified them to better suit their needs and aesthetics. In the case of paper looking glasses, Native peoples modified mirrors when they sewed them onto clothing or other items. It was also a common practice for Native peoples to remove mirrors from European-made casings and place them into Native-made holders (figs. 3.5, 3.6). During his 1832 stay with the Mandan along the Missouri River, German native Maximilian, Prince of Wied, gave a detailed description of modifications: once the traders had sold the looking glasses (which in this case came in paper cases), the consumer immediately took the glass from its frame and placed it in "a solid wooden frame," which was "attached to the wrist by a red ribbon or a leather strap." The frame was "often painted red, or with

Figure 3.6. Wooden casing for trade mirror, excavated near Mandan, Minnesota, 1800–1850. Courtesy of the Minnesota Historical Society.

stripes of different colours, with footsteps of bears or buffaloes carved on it." Some of the wooden frames Native peoples used to enclose European mirrors were quite large and are known as "mirror boards" and might be put to ceremonial uses, which will be discussed in chapter 6. Maximilian noted one such mirror board that was "divided at one end like a boot-jack, and ornamented with brass nails, ribbons, pieces of skin and feathers." Others might take the basic shape of the bootjack but with less decoration (fig. 3.7). The portrait of Upsichtä, a Mandan man whom Karl Bodmer painted in the early 1830s along the Missouri River, provides another example of how Native Americans modified European trade mirrors, in this case by integrating the looking glass into a grouping of eagle feathers, secured by a leather cord. Placing the mirror on the inside of the feathers enabled Upsichtä to catch his reflection in it easily whenever he desired to do so (fig. 3.8).[55] In the process of all of these modifications, Native peoples shifted the object away from its European roots and positioned it as their own.

The evidence shows a robust distribution of mirrors in Native societies—from Pynchon's trade records for the Connecticut River Valley to the Hudson's Bay Company in the Canadian subarctic and the American Fur Company in the western region of the United States. We also know that Native peoples incorporated mirrors on their own terms by creating a demand for paper looking glasses and by framing European glass to meet their own needs and desires. Although it is not possible to quantify mirror distribution in most Native populations, the scores of travelers' and traders' accounts, to which we will return in later chapters, strongly support the idea presented here that looking glasses were well distributed throughout Native societies by the mid-nineteenth century, and much earlier in some places.

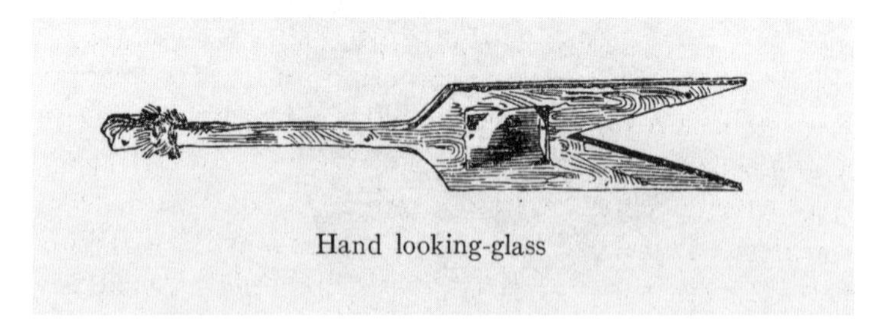

Figure 3.7. "Hand looking-glass." From *Early Western Travels, 1748–1846*, ed. Reuben Gold Thwaites, vol. 23, pt. 2 of *Maximilian, Prince of Wied's Travels in the Interior of North America, 1832–1834* (Cleveland, OH: Arthur H. Clark, 1906), 267.

Figure 3.8. Upsichtä, Mandan Man (1834), by Karl Bodmer (Swiss, 1809–93), watercolor on paper. Joslyn Art Museum, Omaha, Nebraska. Gift of the Enron Art Foundation, 1986.49.263.

European Americans

The following discussion is based on a sample of more than 2,300 probate inventories, beginning in the mid-seventeenth century and recorded through the mid-nineteenth century, that document changing patterns of mirror ownership among people of European descent in North America.[56] Of these inventories, 64 percent come from the seventeenth- through the nineteenth-century records of seven counties in Massachusetts, South Carolina, New York, and Kentucky to provide a sample that captures change over time and includes different geographic regions as well as urban and rural populations. The records from these seven counties were sampled beginning with the earliest available probate inventories. Based on the years of available inventories for each county, I selected sets of years from which to collect inventories and left temporal gaps between each set of inventories. For the purposes of this discussion a "set of inventories" refers to all the inventories from one county in a continuous span of years. For example, the sets of inventories from Hampshire County, Massachusetts, included in this study are 1678–97, 1709–15, 1741–45, 1774, and 1813–16. The number of years included in any one set of inventories varies considerably because it was sometimes necessary to look at several years in order to get a robust number of inventories to analyze. The sets of inventories do not correspond by years between counties because counties were established at different times, and I started with the earliest years for which there were probate inventories in each county. The remaining 36 percent of inventories in this study were drawn from the work of Alice Hanson Jones, an economic historian who published inventories from 1774 from ten of the original thirteen British North American colonies. This second set of inventories allows a snapshot across twenty-five counties in New England, the Middle Colonies, and the South of mirror ownership in 1774.[57] For each set of inventories, the median estate value was identified and the estates were then divided into quartiles, from the poorest to the wealthiest: lowest quartile, lower middle quartile, upper middle quartile, and upper quartile.

Three counties in the study—Suffolk County, in Massachusetts; New York County, in New York; and Charleston County, in South Carolina, respectively—contained the major urban centers of Boston, New York, and Charleston. In 1760 they were three of the four largest cities in North America.[58] While the city of Charleston's population did not keep up with the burgeoning population of New York or Boston after independence, it continued to be an important southern urban center and port city.[59] Four rural counties were also chosen for this study: Hampshire County, Massachusetts, settled in the mid-seventeenth century; Spartanburg County, South Carolina, and Fayette County, Kentucky, both settled at the close of the eighteenth century; and Greene County, New York, established in 1800.

In European American households, accurately reflective glass mirrors were first acquired primarily by the wealthy and were concentrated in urban areas. In Suffolk County, Massachusetts, where Boston is located, between 1660 and 1662 mirror ownership in the two quartiles below the median averaged 13 percent, whereas above the median, ownership reached an average of 31 percent. By 1695 to 1697, mirror ownership among the upper quartile of citizens of Suffolk County was 83 percent, while ownership among the other three quartiles averaged 32 percent. The pattern was similar in Charleston County (SC), where the city of Charleston is located. Settled only in 1670, Charleston County had a small number of inventories available from around the end of the century, but in twenty-seven inventories from 1693 to 1718, mirror ownership in the upper quartile reached 71 percent but among the other three quartiles averaged 25 percent.[60] Mirror ownership in rural Hampshire County, Massachusetts, lagged behind the more urban areas.[61] Hampshire County, which was established by the colony in 1662, encompassed the western third of the colony, stretching from its northern to southern borders around the fertile Connecticut River Valley. In Hampshire County the overall rate of looking-glass ownership between 1678 and 1697 was 13 percent, with only 23 percent of the wealthiest citizens owning a looking glass and none of the county's poorest residences possessing one. In the wealthiest quartile, however, mirror ownership in Hampshire County spiked to 71 percent by the period 1709 to 1715, while among the other three quartiles it rose to an average of 29 percent.[62]

It was not only that mirrors were owned primarily by the elite during the early period of European settlement in North America; the wealthiest members of colonial American society also presumed the mirror to be one of many luxury items that belonged solely in their purview. This attitude is displayed in an account from the Scottish physician Alexander Hamilton's journey from Maryland to Maine in 1744. Along the way Hamilton and his traveling companion, a Mr. Milne, encountered a family of nine who lived in a plain cottage. To his surprise, Milne observed "severall superfluous things which showed an inclination to finery in these poor people, such as a looking glass with a painted frame, half a dozen pewter spoons and as many plates, old and wore out but bright and clean, a set of stone tea dishes, and a tea pot." Milne insisted that the Stanesprings should sell these items "to buy wool to make yarn" and proposed several less expensive items they could use instead of these luxuries, such as wooden plates and spoons. But what if the family still wanted to be able to see their reflections? Milne suggested "a little water in a wooden pail might serve for a looking glass."[63]

The Stanesprings were not alone in their ownership of what the elite presumed to be luxury items. Despite Milne's insistence that it was inappropriate to own finery when in need of necessities for living, many eighteenth-century Americans began to reject restrictions that others attempted to place on their consumption. As the historian T. H. Breen has argued, over the course of the eighteenth century, as British imports became increasingly available in the North American colonies, especially after 1740, less wealthy whites had increasingly varied options of things that they might purchase. They showed increasingly little regard for purchasing what the wealthy deemed appropriate to their station. The *South-Carolina Gazette* reported this exchange in 1736: "A woman has entered a small shop, and every time she inquires about the cost of a certain object, the proprietor delivers a gratuitous sermon about the vanity of such goods. Finally, a little out of patience, the woman declares, 'Yes, Sir, but I did not ask you the Virtues of it, I ask'd you the Price.'" This woman chose to consume based on what she desired and could afford, not on what anyone else deemed to be appropriate for her.[64]

What did continue to set wealthy whites apart from everyone else was the type of mirrors they purchased and the number of mirrors they owned. When Boston lawyer Josiah Quincy Jr. visited Charleston, South Carolina, in 1773, he declared that "in grandeur, splendour of building, decorations, equipages, numbers, commerce, shipping, and indeed in almost every thing . . . [Charleston] far surpasses all I ever saw, or ever expected to see, in America." While Quincy had certainly seen extensive material opulence in his hometown, he reserved highest praise for Charleston. Visiting the Georgian-style home that had been built between 1765 and 1769 for Charleston merchant Miles Brewton, Quincy observed "the grandest hall" he had "ever beheld." It was decorated with "azure blue satin window curtains, rich blue paper with gilt, mashee borders, most elegant pictures, excessive grand and costly looking glasses etc." Another Charleston home, completed in 1808 in the Federal style for a merchant, Nathaniel Russell, not only had the opulent wall-hanging and tabletop looking glasses common by then in the homes of the wealthy but also boasted two stately floor-to-ceiling mirrored panels in an upstairs drawing room that mimicked the four eighteen-paned windows opposite them (fig. 3.9). Standing in front of a grand mirror in the public spaces of these fine Charleston homes, an observer would have seen an image not only of himself or herself but also of various luxury items and, perhaps, other notable people, captured within the reflection. Moreover, the entire tableau would have been enclosed within the mirror's elegant frame. As Quincy's observations after his visit to Miles Brewton's home suggest, wealthy white Americans showcased these grand mirrors, and the scene they helped to establish, as one of the myriad material ways to proclaim their owners' wealth and prestige.[65]

Figure 3.9. Oval drawing room, Nathaniel Russell House. Photograph by Rick McKee. Courtesy of Historic Charleston Foundation.

That the wealthy intended the grand mirrors in their public spaces to make a statement about their material success and elite status can be seen by contrasting those mirrors with the smaller, plainer ones that the wealthy commonly chose for more private spaces. Josiah Baker's plantation house on the Ashley River, just inland from Charleston, South Carolina, had a "hall" and "chambers" according to its 1743 appraisal. The public space of the hall included two looking glasses and one pair of sconces valued at £12. In the private chambers, there were two looking glasses: an "Old Dressing table & Glass" valued at £5 and a "Chamber table & Looking glass" valued at £1.15s. That same year, the appraisers who visited the home of James St. John, Esquire, in Charleston found a "Pier Glass and Sconces" valued at £15 in the front hall, one in an upstairs chamber valued at £5, and, in the "Back Room," a mirror worth only £1.[66]

Ownership of multiple mirrors also distinguished the wealthy from the common folk. Whereas people of more modest means often acquired only one mirror, multiple looking glasses were common in the homes of the elite. In the estates probated in Suffolk County, Massachusetts, in 1813, the ten wealthiest homes all included a mirror. Nine of those ten owned more than one, and in four of these estates there were more than five mirrors. The single wealthiest person inventoried that year, Thomas Amory, owned ten mirrors. Among the ten poorest people inventoried that year in Suffolk County, three owned a mirror, but no one had more than one.[67]

While only the wealthy could afford multiple mirrors and the opulent looking glasses showcased in the public spaces of their homes, modest mirrors became more common across all class levels in eighteenth-century European American homes. In Charleston County, South Carolina, the average rate of mirror ownership across all quartiles averaged 49 percent from 1739 to 1743 (up from 37 percent from 1678 to 1721) with the lowest and lower-middle quartiles having rates of 27 percent and 40 percent; in the upper-middle and upper quartiles, ownership rates reached 63 percent and 75 percent. In Suffolk County, Massachusetts, by 1742 and 1743, the average rate of mirror ownership for all quartiles reached 68 percent (up from 45 percent during the 1695–97 period), with the lowest percentage of ownership being 58 percent in both the lowest quartile and the upper-middle quartile. In Hampshire County, Massachusetts, mirror ownership between 1741 and 1745 averaged 50 percent (up from 40 percent during the 1709–15 period), with rates of ownership in the lowest quartile at only 15 percent, but in the lower-middle quartile ownership rates reached 42 percent. In New York County, between 1746 and 1755, mirror ownership across all quartiles averaged 74 percent, with the lowest percentage of ownership being 43 percent in the lowest and lower-middle quartiles.[68] As looking glasses became more common among people at the lower end of the economic spectrum, it is important to remember that the purchase of a looking glass could represent making a difficult choice about how to spend one's money; this purchase might also represent the expenditure of money that had been saved over a significant period of time. In other words, these purchases would have been considered carefully by the buyers, whose actions indicate the importance they placed on this item.

Eighteenth-century craftsmen, merchants, and shopkeepers stocked looking glasses in a wide range of prices to meet the demand of this disparate customer base. Specialty stores run by craftsmen who described themselves as carvers or gilders sold the most expensive mirrors.[69] These shops carried a wide range of looking-glass styles, as seen in this 1763 advertisement from Edward Weyman of Charleston, South Carolina: "A very large and neat assortment of looking-glasses, the best ever imported into this province, consisting of mahogany and walnut chimney glasses, pier and sconce glasses of most sizes and fashions, dressing glasses, shaving glasses, and pocket glasses in painted frames" (fig. 3.10).[70] While Weyman's advertisement described looking glasses that had been imported and already framed, he also had glass plates that could be fit into a customer's preexisting frame. Craftsmen also made frames for looking-glass plates that had been imported unframed. Another Charleston craftsman, George Stattler, boasted that, with his experience in England and well-established connections there, he could produce framed looking glasses "equal to any imported." One did not have to live in or travel to the city to acquire one of these fine glasses. Merchants offered to supply "country" as well as city orders, with promises that the glasses would be "packed up with the greatest Safety."[71]

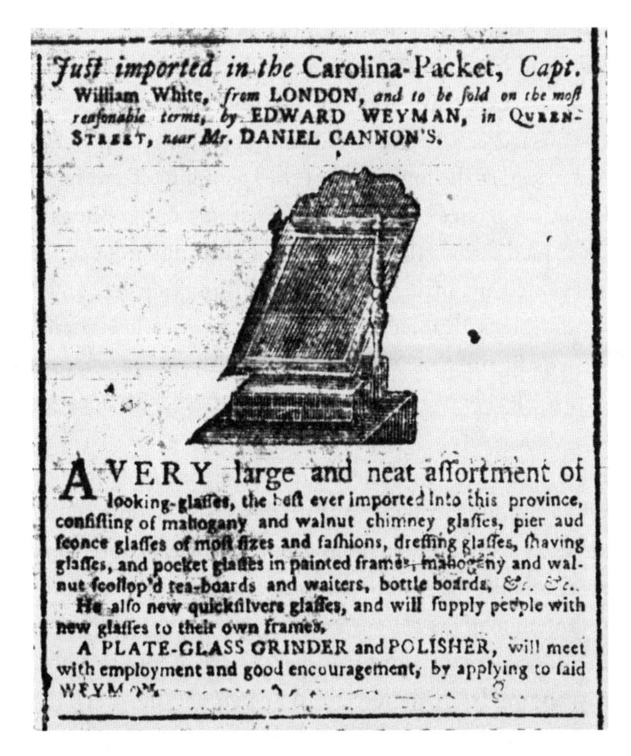

Figure 3.10. Edward Weyman advertisement. *South-Carolina Gazette*, Nov. 5–12, 1763, 1.

Both Weyman and a Philadelphia craftsman, John Elliott, also repaired looking glasses that did not produce accurate reflections. At his Philadelphia "Looking-glass Store," Elliott promised customers that he could "undertake to cure any English Looking glass, that shews the face either too long or too broad, or any other way distorted."[72] The mirrors that Elliott repaired may have suffered from imperfections in the flatness of the glass or the reflective coating that caused these visual distortions. Similarly, Weyman advertised that he could remove "Spots and Stains which prevent them from regularly reflecting the Objects placed before them."[73] That both Elliott and Weyman chose to advertise this service suggests that it was a widespread problem that could bring them revenue. Their readiness to undertake this work for their customers also speaks to the expense of larger mirrors, which would have been cheaper to repair than to replace in this era.

These specialized stores and repair shops might offer less-expensive looking glasses as well. Elliott offered "a large and neat Assortment of Looking-glasses of most Sorts, Sizes, and Fashions," at a wide range of prices, "from one shilling to fifteen Pounds and upwards, per Glass."[74] But general stores were the more likely kind of establishment from which people of modest means acquired smaller, less-expensive looking glasses. Between 1739 and 1743 the three Charleston-area stores that entered the probate process all carried looking glasses priced at the lower end of the spectrum. Together they stocked: "three small looking glasses" valued together at just under one pound, twenty-seven mirrors valued at a little more than 4 shillings a piece, seven "small" looking glasses valued at a little more than 2 shillings a piece, and sixty looking glasses of "sundry sizes," valued on average at a little less than 3 shillings a piece.[75] While costly mirrors allowed the wealthy to proclaim their status, mirrors at all price levels provided white consumers with access to their own reflections.

In her study of colonial wealth near the end of the eighteenth century, Alice Hanson Jones compiled probate inventories from a sample of twenty-five colonies located in the three main regions of British settlement in North America: New England, the Middle Colonies, and the South from 1774. These inventories reveal an average rate of looking-glass ownership in North America at this time of 60 percent, with the overall rate of looking-glass ownership in each colony ranging from 49 percent to 75 percent (fig. 3.11). In these inventories, the lowest quartile had a mirror ownership rate of 36 percent, while in the lower middle quartile mirrors could be found in 55 percent of households.

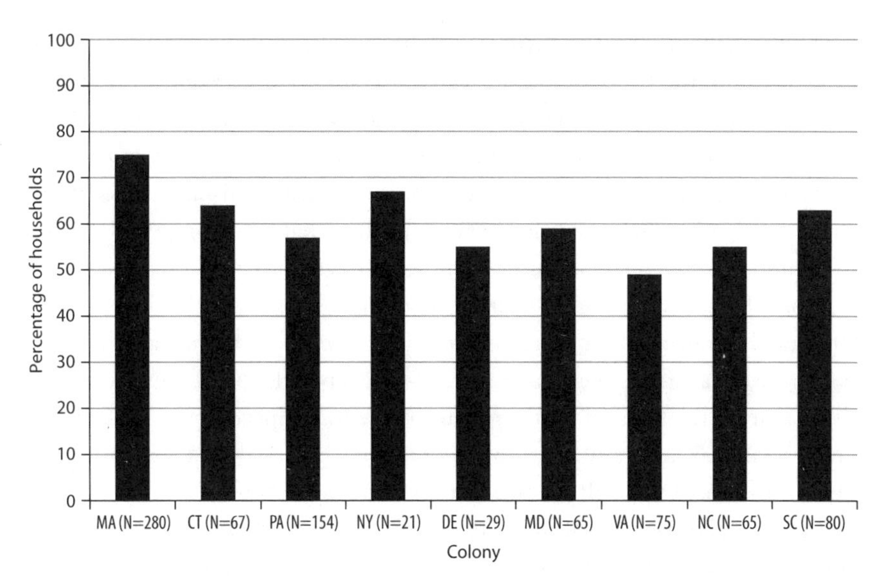

Figure 3.11. Percentage of households that owned a mirror by colony, 1774. Data compiled from Alice Hanson Jones, *American Colonial Wealth: Documents and Methods*, 3 vols. (New York: Arno, 1977).

In the two quartiles above the median, mirror ownership reached 68 percent in the upper middle quartile and 85 percent in the upper quartile.[76] By the end of the eighteenth century, then, white Americans not only had access through several avenues to acquire looking glasses, but a majority of them—or at least of those who had been probated—had done so.

In the nineteenth century, mirror ownership among whites continued to increase in the individual counties that are the focus of this study. In 1813 Suffolk County, Massachusetts, mirror ownership across all quartiles averaged 84 percent, with the lowest rate of ownership at 67 percent in the lowest quartile. Although Suffolk County had discontinued the practice of inventorying individual property by midcentury, Plymouth County, located just to the south of Suffolk, continued to collect this information. Here, in 1850, mirror ownership across all quartiles averaged 72 percent, with the lowest rate of ownership at 53 percent in the lowest quartile.[77] In 1823 Charleston, South Carolina, ownership at all quartiles averaged 78 percent, with the lowest rate of ownership at 53 percent in the lowest quartile. In Hampshire County, Massachusetts, from 1813 to 1816, ownership at all quartiles averaged 67 percent, with the lowest rate of ownership at 55 percent in both the lowest and the upper-middle quartiles.[78]

The nineteenth century also allows us to look at white settlement as it expanded into the backcountry and moved west into new regions. Located in the upstate of South Carolina, Spartanburg County was founded in 1785. Travelers' impressions of life in early nineteenth-century Spartanburg County have portrayed a rough existence. Jonathan Mason described his accommodations there as being in "a miserable log-house" that had "but one room, two beds full of vermin, and not a single thing of any kind to eat or drink." A similar story was told of the "lower class" in Spartanburg, whose houses were "more wretched than can be conceived." The 1826 *Statistics of South Carolina* indicated that the county seat—also known as Spartanburg—had a population of only eight hundred. During the first half of the nineteenth century Spartanburg County had an agricultural economy that "was dominated by small and medium-sized farms." In 1860 it was still the third-poorest county in South Carolina.[79]

Probate records taken between 1788 and 1810 show the average rate of mirror ownership across all quartiles in this earliest period of Spartanburg County's history to be 35 percent. In the lowest quartile of homes, 11 percent owned a mirror, but none of the households that fell in the lower-middle quartile (the other group below the median) possessed one. But the upper-middle quartile already had a 56 percent ownership rate and the upper quartile a 75 percent ownership rate. By 1835 to 1840, mirror ownership had increased in all quartiles in Spartanburg to an average of 48 percent, with ownership in both quartiles below the median at 27 percent. From 1854 to 1855, mirror ownership rose again across all quartiles to an average of 58 percent: rates of ownership above the median were 85 percent in the upper quartile and 71 percent in the upper-middle quartile; below the median, ownership rates were 31 percent in the lowest quartile and 43 percent in the lower-middle quartile.[80] Many Spartanburg County residents chose to acquire mirrors during the first half of the nineteenth century.

Evidence from the five Spartanburg County store inventories that entered the probate process during this period also suggests that looking glasses were readily available and also indicates that the stock of this item increased over time. Two stores were inventoried between 1808 and 1809. One store had six looking glasses in stock (valued together at $1.00), although the other did not have any available for sale. Two stores inventoried between 1835 and 1840 had looking glasses in stock: one of these stores had a "lot" of looking glasses valued at $9.00; the other had eight gilt looking glasses worth $7.00. The final store inventory was taken in 1855. Here Ezekiel Webster had thirteen looking glasses for sale—nine labeled as mirrors and four as looking glasses. The mirrors were valued at $1.00 each; one of the looking glasses was worth $0.37, while the other three were worth only $0.09 apiece. That more expensive looking glasses appear later in the century also suggests that larger, more decorative looking glasses used in part for display (especially those labeled "gilt") may have replaced small, plain ones in some households.[81]

Greene County, New York, was established in 1800, located along the Hudson River, roughly 125 miles north of New York City. While New York City was establishing itself as "the nation's economic capital," the majority of those living in the mid–Hudson Valley, including those in Greene County, still "struggled to make ends meet" in the years after the American Revolution. In contrast to Spartanburg County, South Carolina, Greene County, New York, already had thirteen thousand settlers at the time it was designated a county. But the rural character of this area persisted throughout the antebellum period. Even by the mid-nineteenth century, "four-fifths of the mid–Hudson Valley's inhabitants—artisans and shopkeepers no less than farmers—worked and lived in neighborhoods where agriculture provided the large majority of people with a livelihood."[82] Between 1800 and 1855, the overall rate of mirror ownership in Greene County grew slowly, from 56 percent between 1800 and 1810 to 61 percent between 1821 and 1830, finally reaching 66 percent between 1841 and 1850.[83]

Fayette County, Kentucky, was included in this study as a representative of an early American western settlement. Kentucky was "settled by yeoman farmers, small planters, and their slaves from western Virginia, Pennsylvania, Maryland, and northwest Carolina during the 1770s and 1780s." By the time of the first United States Census in 1790, Kentucky had only five towns. Lexington, Fayette County's seat, was the largest town in 1790, with a population of 834. Kentucky became an important area for agricultural and livestock production. Lexington, which quickly developed into a processing and manufacturing center for these products, despite being landlocked, would maintain its status as the largest city in Kentucky for thirty years, growing to a population of 5,279 in 1820 and 6,026 in 1830. Lexington, commonly referred to as the "Athens of the West," also established itself as an important early cultural center in the American West.[84]

In "The Consumer Frontier: Household Consumption in Early Kentucky," historian Elizabeth Perkins studied Lexington's Fayette County and Louisville's Jefferson County, which, because of its convenient location on the Ohio

River, surpassed Lexington in population by 1830. Perkins challenged the assumption that "the Appalachians formed a barrier only slowly penetrated by merchants and consumer goods." She found that in addition to people bringing consumer goods with them when they moved west, there were also stores in the area from the outset of European settlement. Looking-glass ownership rates confirm Perkins's findings. Looking-glass ownership rates in Fayette County across all quartiles was already 56 percent in 1803; the rate of mirror ownership increased to 64 percent from 1813 to 1834 and finally to 73 percent in 1853. Between 1803 and 1853, looking-glass ownership fell below 50 percent only two times: in the lowest quartile in 1803 (20 percent) and in the lower-middle quartile in 1813 to 1814 (39 percent).[85]

While there certainly remained, by the 1850s, individual households in which inventory-takers did not document the presence of a mirror, nineteenth-century white Americans had increasing opportunities to come to know their mirror selves.[86] Even members of households that did not yet own a mirror would have been likely to have a friend or family members of the same class status who did possess one that they could see themselves in at least occasionally. Many would have agreed with Charles Dickens's assertion in *Our Mutual Friend* (published in the United States in serial form in *Harper's New Monthly Magazine* from 1864 to 1865) that one of the "consequences of being poor" by this time was *not* having to make do without a looking glass. For "a girl with a really fine head of hair," being poor now meant "having to do it by one flat candle and a few inches of looking-glass."[87] In other words, people might not own the mirror they desired (in this case, one large enough for a young girl to easily see herself as she fixed her hair), but they likely had access to one when needed. Even in the backcountry of South Carolina, upstate New York, and the early western frontier in Kentucky, by the 1850s, mirror ownership averaged 58 percent (South Carolina), 66 percent (New York), and 73 percent (Kentucky). In longer-settled and more urban regions, average rates of mirror ownership across all quartiles in the nineteenth century consistently reached more than two-thirds of the population: in Suffolk County, Massachusetts, 84 percent by 1813–14; in Plymouth County, Massachusetts, 72 percent by 1850; in Hampshire County, Massachusetts, 67 percent by 1813–16; and in Charleston County, South Carolina, 78 percent by 1823. Nineteenth-century white Americans at all class levels had become intimately familiar with their mirror selves.

Tracing patterns of ownership of looking glasses across early North America reveals much about the distribution of this particular item of material culture among people of African, Native, and European descent. But understanding distribution is only the first step in apprehending the meaning and significance of any material culture item. We must also ask about how objects were used, by whom, and what meanings were associated with them. This is perhaps especially important in understanding the uniquely visual looking glass, an object that provided men and women with access to critical knowledge about themselves but that also linked them back in time to magic and ritual practices about reflection.

Reliable Mirrors and Troubling Visions

Nineteenth-Century White Understandings of Sight

As soon as Abraham Lincoln was nominated for the office of the president of the United States of America by the Republican National Convention in 1860, he went home to tell his wife the news. In the upstairs bedroom where Lincoln found her, he stretched out on a lounge, opposite "a bureau, with a swinging glass upon it." As Lincoln told it, looking into that glass, "I saw myself reflected, nearly at full length, but my face, I noticed, had *two* separate and distinct images, the tip of the nose of one being about three inches from the tip of the nose of the other." When he stood up and looked again in the glass, everything appeared normal. But, "on lying down again," in an attempt to duplicate what he saw, Lincoln found that he "saw it a second time—plainer, if possible, than before." He admitted that he was "a little bothered, perhaps startled," by the "ghost" he had seen. Lincoln also noticed that one of the images "was a little paler, say five shades, than the other." Lincoln admitted that while he was quickly able "nearly" to forget about these incidents, he was left with a lingering "pang, as though something uncomfortable had happened." Mrs. Lincoln believed that this was a "sign" that the president "was to be elected to a second term of office, and that the paleness of one of the faces was an omen that [he] should not see life through the last term." The person who told this story brushed off this explanation and suggested that Lincoln himself "saw nothing in all this but an optical illusion" and that it was in "entire consistency with the laws of nature." But even the narrator acknowledged that "the flavor of superstition which hangs about every man's composition made [Lincoln] wish that he had never seen it."[1]

By the nineteenth century, white men and women relied daily on mirrors and trusted them to provide key visual information about themselves. That the mirror image startled Lincoln on that day in 1860 suggests that he, too, had a lifetime of encounters with his mirror self with which to compare this most unusual one.[2] Over the course of the nineteenth century, theoretical and scientific explanations emerged that explained optical illusions, like what Lincoln saw that day, which is why the person who recounted the story understood what happened as having "entire consistency with the laws of nature."[3] But around midcentury, encounters before the looking glass began to appear in the historical record that captured moments—like Lincoln's—when men and women seem to question what their eyes revealed to them about their mirror selves. This chapter explores both the ways in which white men and women trusted mirrors to provide accurate representations of the world and seeks to contextualize the concerns that arose midcentury about whether what was encountered in the mirror could be trusted to be an accurate representation of reality.

The Reliable Mirror

Nineteenth-century whites used looking glasses as critical and trusted visual sources of information about the "public self." As the sociologist Erving Goffman described it, the public self is that version of the self that an individual puts forth "in the presence of others" to "convey an impression [to them] which it is in his interests to convey." A key component of the public self is the "personal front," which comprised an individual's "office or rank; clothing; sex, age, and racial characteristics; size and looks; posture; speech patterns; facial expressions; bodily gestures; and the like."[4] Accurately reflective glass mirrors enabled whites to monitor their public selves and, as a result, became trusted visual tools in gaining knowledge about, and exerting mastery over, many aspects of the personal front. In North America, mirrors were embraced by both men and women for this key information about the self that they provided.

Both men and women used the mirror to better know how they looked, and they acquired key self-knowledge in the process. White men and women trusted what the mirror told them about themselves and used the knowledge they gained in encounters with their mirror selves in constructing their identities. But the meaning of this work differed significantly by gender. For white women the mirror also acted as a technology in which they could monitor their public selves and attempt to achieve a personal front that would be attractive to men. In addition to using mirrors to prepare their personal fronts and to gain knowledge about themselves, white women also used mirrors to assess the physical attractiveness that was key to how society valued them. A young woman described in a mid-nineteenth-century story from *Harper's Weekly* explained perfectly the connection between a woman's appearance and her value when she sat down before a looking glass "to study a little my face and my prospects."[5] This woman understood that the attractiveness of her face—which mirrors allowed her to be intimately familiar with and which had come to visually represent peoples' identities—was directly related to her "prospects," how successful she could be in life. While white women's attractiveness had certainly been important before the advent of accurately reflective looking glasses, mirrors enabled women to become monitors and judges of their own appearance in ways previously impossible.

The mirror came to be a substitute for men, and other women, who had long judged visually whether a woman's appearance was acceptable. In her conduct book from the mid-eighteenth century, *The Female Spectator*, English author Eliza Haywood had described this male gaze. Women were preoccupied with their appearance, Haywood argued, because men regularly surveyed women as though they had set "themselves up for Sale." Men considered themselves "Buyers" at these sales and "measure us with their Eyes from Head to Foot." Because this atmosphere of acquisition permeated encounters between men and women, men had come to expect a perfect product in the women they chose. There were few women, Haywood argued, who had not encountered a man who would "pass by her with a contemptuous Toss" because her appearance did not meet his expectations. With men regularly gazing on them in such a "scrutinous Manner" from "Head to Foot," white women readily embraced the opportunity looking glasses offered to view themselves to better know how they looked and to prepare their public selves. In doing so, white women used the mirror to turn the male gaze upon themselves. Women also assessed and judged other women as well, but those critical glances often similarly enforced the standard of beauty sanctioned by men.[6]

As a substitute for the male gaze, white women wrote about their mirror selves in ways that gave agency to the mirror, which was, essentially, playing the role of a man in these encounters with the looking glass. A member of Philadelphia's elite, Anne Home (Nancy) Shippen Livingston (1763–1841) recorded in a diary entry from May 1783 how she was pleased when her glass "told" her she "look'd well" in her pink dress and petticoat. When Eleanor Parke Custis Lewis (1779–1852)—George Washington's step-granddaughter—was traveling in the hopes of improving her health in 1822, she became discouraged in hearing from her friends that she was not looking any better. More than their word, however, she trusted her mirror, which confirmed their conclusion. Her glass, she wrote, "tells me I never look'd less healthy." In 1849 Elizabeth Blackwell (1821–1910) became the first woman to graduate from an American medical school. A few years before embarking on this career path, Blackwell accepted a job at a girls' school in Henderson, Kentucky. Blackwell wrote in a letter home that upon her arrival in Kentucky, she "did not care one straw what was thought of my personal appearance." Yet Blackwell was only able to maintain this stance before she had established connections with Kentuckians, including some potential suitors (all of whom she rejected). Blackwell admitted that "now I sometimes dress for others, and feel a slight satisfaction if the glass tells me I shall not scare people." Sometimes women believed what the mirror told them even when other evidence contradicted it. In 1865, New York native Annie Van Ness (b. 1848) wrote that although she felt like she had recently "lost ten pounds," she was disappointed that "when I look in the glass I look as fat as ever."[7] The mirror was an especially trusted source of information about their appearance and bodies for white women.

Nineteenth-century white women knew the general outline of what they should seek to achieve in their mirror selves. As former mill worker from Beverly, Massachusetts, Lucy Larcom (1824–93) described it in the memoir of her early life, this "ladyfied standard" of women's beauty was exemplified by women who were "pale and pensive-looking," with "high white foreheads" and "cheeks of a perfect oval." An 1858 illustration from *Harper's New Monthly Magazine* favorably depicted "Civilization" as an elegantly dressed woman who typified these standards. It was no accident that this emblem of refined beauty stood in front of a large mirror (fig. 4.1).[8]

Some women observed conformity to these standards of beauty when they encountered their mirror selves. A young Louisa May Alcott (1832–88), the author and reformer, born in Germantown, Pennsylvania, admitted that as a young woman, when she looked in the glass, she had to "try to keep down vanity about my long hair, my well-shaped head, and my good nose." But women more frequently expressed displeasure at how their mirror selves stacked up against beauty standards. Future first lady Abigail Adams (1744–1818) found herself living in the 1780s in an elegant mansion in Auteuil, outside of Paris. The house was replete with mirrors. Adams believed she was "rather clumsy and by no means an elegant figure," and she disliked being in rooms of this house where she saw these qualities of herself "multiplied" by looking glasses. Lucy Larcom admitted that she "did not like to look at my own face in

Figure 4.1. "Civilization," *Harper's New Monthly Magazine*, Jan. 1858, 177. Courtesy of Cornell University Library, Making of America Digital Collection.

a mirror" because it was "round" and "ruddy," and no matter how long she practiced before her mirror, she "could not lengthen it by puckering down" her mouth.[9]

Women saw their ability to conform to societal standards of beauty slipping away as they observed the aging process in the mirror. In 1809, when she was sixty-five, Abigail Adams suffered from an illness that "swelled and inflamed" her face almost to the point of blindness. During her recovery, Adams told her sister that she wanted to take "a little journey" but regretted that "as years and infirmities increase, my courage and enterprise diminish." Adams expressed her frustration that age was "dark and unlovely." Looking in the mirror, she recalled the story of Zeuxis, a painter, who reportedly "died of laughing at a comical picture he had made of an old woman." Adams concluded that "if our glass flatters us in youth, it tells us truths in Age." Esther Hill Hawks, a "graduate of the New England Female Medical College" who traveled south to provide medical care during the Civil War and kept a diary of her experiences, marked her thirty-first birthday with this observation: "if *mirrors* could be abolished, I could not believe that I am *growing old*—but with these faithful friends present, before me I cannot forget the rolling years even if I would." Even Louisa May Alcott, who had admired her mirror self as a young woman, wrote around 1870, at age thirty-eight, that her mirror showed "how fat and old I was getting."[10]

Although women spent a great deal of time considering their mirror selves and seeking to achieve the societal standard of beauty, they were discouraged from admitting that their looks were anything but effortless. Mary Willard (1843–62), the younger sister of famed temperance and women's rights activist Frances Willard, admitted in 1861—just a year before her death at nineteen—whenever she heard the sounds of approaching visitors, she would always look in the glass "to see if the toilet is all right." Yet she did not want her company to know what she was doing. To that end she admitted that she would "peep slyly into the glass" lest anyone observe what she was doing. Already caught in conversation with others at a party near the end of the Civil War, Georgia native Eliza Andrews (1840–1931) suddenly became worried about her appearance. She excused herself "and slipped off upstairs to get a look at myself in the glass" in private. Minimizing their reliance on the mirror self when in public downplayed whatever vanity women did have and enabled white women's beauty to seem natural and effortless.[11]

In private, mirrors also gave women an opportunity to look at themselves for a long time. Usually such encounters with the mirror self focused on features of the body that could be scrutinized and enhanced before the looking glass. But looks alone were not all that mattered if women were to live up to the "ladyfied standard" of refinement. They also had to have pleasant countenances and personalities. The mirror could help with preparing this aspect of the public self as well. Sarah Morgan (1842–1909), the daughter of a prominent Baton Rouge attorney, sometimes found herself in such a bad mood that she needed someone to whom she could "vent" her feelings. She worried that these feelings would be inappropriately expressed before her family and friends and advised that "if ill humor must have a vent, make faces at your looking glass" so that when you "enter the family circle," you would be able to have "a pleasant word for all" and contribute "to the common happiness." Frequently, she followed her own advice, once returning to her room and making "the most hideous faces at myself" after upsetting her brother and, on another occasion, "making faces at myself" and "whisper[ing] as though some one was near." In conversing with the mirror self, Sarah Morgan exorcised emotions that she believed should not be expressed in public yet needed an audience, even if that audience consisted only of herself. In addition to being able to aid in the curling of hair or the proper fitting of a dress, then, the mirror could also provide a place where a white woman could be decidedly unladylike with the satisfaction of performing before an audience who would never reveal that she had engaged in such inappropriate behavior.[12]

Antebellum white women could seek out their mirror selves in a wide range of shapes and sizes of looking glass: pocket glasses, small hand mirrors, tabletop glasses meant to be set on top of another piece of furniture, dressing tables that were elaborate pieces of furniture themselves, and full-length mirrors (figs. 4.2, 4.3, 4.4, 4.5). Dressing tables were designed specifically for women's daily preparations of their public selves. Women sat, or perhaps sometimes stood (see fig. 4.1), before these tables, where they could study themselves in the mirror and prepare themselves for the day. Some of the tabletop and

Figure 4.2. Hand mirror, 1790–1800, England; mahogany, beech, paint. Museum purchase, 1954.16.11. Courtesy of Winterthur Museum.

Figure 4.3. Dressing glass by Stephen Badlam Jr., 1800–1820, Boston, Massachusetts; mahogany, white pine, glass. Museum purchase, 1958.39. Courtesy of Winterthur Museum.

full-length looking glasses were "swingers"—attached to the stand with a wing nut or spring catch for adjustability so that each user could find precisely the best viewing angle.[13] By 1840, Eliza Leslie—an author of advice books—wrote that mirrors that sat on top of dressing tables were now out of fashion because they "scarcely show more than your head, and are very easily upset." Replacing these tabletop glasses were much larger ones that it was "now customary to fix . . . upon the wall at the back of the table or bureau; suspending it by a double ribbon to a strong hook, and making the string long enough to allow the glass to incline considerably forward, so as to give the persons that look into it a better view of their figures."[14]

To see parts of the body inaccessible before the mirror, some women began to use two mirrors—one held in the hand—to see the back side of their bodies. That same 1840 advice book opined that "in every chamber should be a second

Figure 4.4. Dressing table, 1810–20, New York; mahogany, mahogany veneer, white pine, tulip poplar. Museum purchase, 1965.73. Courtesy of Winterthur Museum.

glass, small and easily moved, to take in your hand for the purpose of looking at the back of your head and neck, after dressing; the large glass being in front." Leslie also encouraged households, in an 1849 publication, to be sure visiting women had the opportunity to survey both sides of themselves before being met by the hostess. Visitors should have access to a dressing table and be provided a few moments of solitude before the visit began. To enable women to catch a glimpse "of the back of her head and neck" as they underwent their final preparations, the author suggested providing either a hair brush "with a mirror on the back" or "a regular hand-mirror, of sufficient size to allow a really *satisfactory* view" when used in conjunction with a larger mirror. Annie Van Ness recorded how her sister Lottie was given a "hand glass"

Figure 4.5. Woman at a Mirror, daguerreotype (1856), by Alexander Hesler. National Gallery of Canada (NGC). Used by permission.

for Christmas in 1870 that enabled her to "look at the back of the hair," which, according to Van Ness, "she was in need of." Leslie and Van Ness would have undoubtedly approved of a looking glass that was advertised in 1866. This "toilet glass" was designed to be hung from the ceiling and used with another mirror to enable an observer to view him- or herself from the front and from behind simultaneously. The company offering this product declared it "indispensable to a ladies' toilet" (fig. 4.6).[15]

Women's daily reliance on having access to mirrors as a trusted source of key visual information about the self—"faithful friends," as Esther Hill Hawks expressed it—can be seen clearly when they were forced to live without one. Although that same 1849 advice book cautioned women never to "set out on a

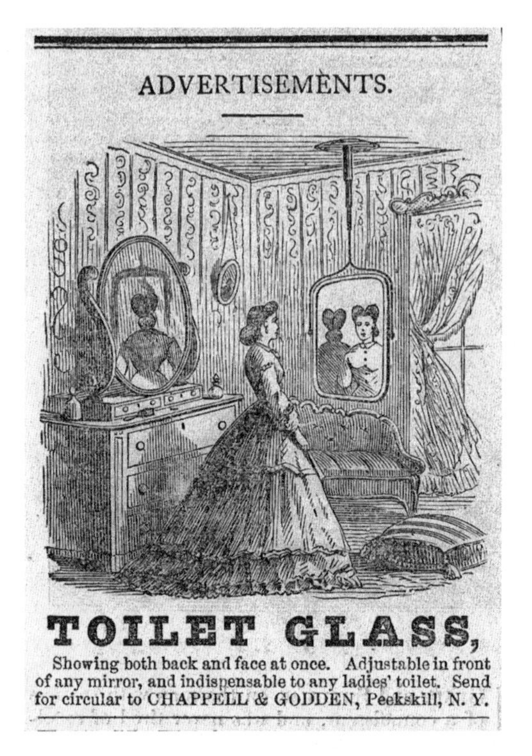

Figure 4.6. Toilet glass advertisement by Chappell and Godden. *Harper's Weekly*, June 18, 1866, 382.

journey unprovided with an oiled-silk bag" for necessities that included a "hair-brush with a mirror at the back," women sometimes found themselves far from home without the aid of a mirror. The anonymous author of *Diary of a Refugee* complained vociferously about her family's accommodations as they traveled to Texas from their Louisiana home during the Civil War. In Laredo, the only accommodation available to them was poorly furnished. The interior was divided into two areas—one for men and one for women—with only a "thin cotton curtain" between them. There was "not a chair or piece of furniture of any kind, nor the tiniest bit of a looking-glass!" Elizabeth Comstock (1815–91), a Quaker minister and abolitionist who traveled during the Civil War "ministering to those in hospitals and prison camps and to refugee slaves," described her frustration with her accommodations in Nashville, where she had stayed in a "comfortless apartment" with "one chair without a seat, one without a back" and "a looking-glass frame without a particle of glass."[16]

Even though women preferred larger mirrors, periods of displacement reminded them that, given the necessity of this item to them, they could make do with a mirror they might have otherwise scorned as much too small and inadequate. When Louise Clappe (1819–1906) first arrived in California in 1851, she mocked the only looking glass available to her there as "one of those which come in paper cases for dolls' houses." She clearly missed "the full-length psyches so almost indispensable to a dressing-room in the States." Soon after her arrival, she left the log cabin for three weeks of travel in the territory. When she returned, Clappe saw her cabin home from a new perspective. What had been worthy of scorn previously now received praise. Clappe was delighted by the "doll's looking-glass"—in which she could scarcely even see her face—"the first time that I had seen it since I left home." New Englander Mary Colby (1806–89) was delighted when in 1852, two years after making the long and treacherous journey by covered wagon to the Oregon Territory, she was finally able to acquire a mirror that she could "se [*sic*] my whole body in." Such displacement might prevent women from consulting the mirror, perhaps even for an extended period of time; these periods of displacement also revealed how much women relied on, and desired, mirrors in their daily lives.[17]

White women not only grew accustomed to seeing themselves in looking glasses; the prominent placement of mirrors in the public and private rooms of many white American homes also gave women extended opportunities to see their loved ones reflected in looking glasses. Mirrors even became memory aids to help women visualize loved ones who were dead or absent from them. In 1779, full of nostalgia, Philadelphian Deborah Norris (1761–1839) wrote to her friend Sally Fisher that she was thinking much of "past scenes" and asked her about the "glass that is in our blue chamber" and whether she ever stood in front of it to remember "thy face by mine, does thee remember how . . . together?" Similarly, Virginian Lucy Breckinridge (1843–65) wrote about Charlie, a young man who was away fighting for the Confederacy during the Civil War. Breckinridge was heartsick for him and wrote about how it was very painful for her to go into the parlor, where she had many fond memories of him. There she saw "the mirrors that have reflected so often his dear face." Now "they reflect nothing but sorrow" when she looked into them.[18] What had once seemed the most ordinary act—catching a glimpse of a loved one in a looking glass—became replete with meaning and significance in the absence of the beloved. Oliver Wendell Holmes wrote of daguerreotypy and photography in 1859, describing the images produced as "the mirror with a memory." This was technically true because the daguerreotype's photographic process affixed a permanent image onto a reflective glass plate (i.e., a mirror). But ordinary mirrors had long served as memory aids, according to Norris and Breckinridge.[19]

Charlie, the young man Lucy Breckinridge missed so desperately, would have had the same opportunities to see young Lucy and his own mirror self reflected in the mirrors of her home. But it is telling that we hear this story from Lucy rather than Charlie. Although the proliferation of mirrors had happened in households occupied by both men and women, it is largely from the women that we learn of the mirror's impact on identity. While men consulted mirrors daily in shaving, grooming, and dressing—and had the same kinds of opportunities to increase their mastery over their public selves as did their female counterparts—men experienced and wrote about their encounters with the mirror self in fundamentally different ways.

In addition to the other mirrors in their homes that men could consult when they desired, by the mid-eighteenth century men had a specific type of mirror made for their use: the shaving glass. Shaving itself had been popularized in the late seventeenth century in Europe when "facial hair began to be seen as uncouth." Although at first shaving was done by barbers, by the mid-eighteenth century, shaving was largely a domestic activity. Domestic shaving glasses could be simple tabletop glasses or elaborate pieces of furniture with drawers and a water basin (fig. 4.7).[20]

Like that of their female counterparts, men's reliance on mirrors can be seen in their complaints when mirrors were either too small or altogether unavailable. Elisha Mitchell (1793–1857), a professor at the University of North Carolina, was unable to find many of the items he needed when he first arrived in Chapel Hill in 1818. He complained to his fiancée that he had bought the "largest looking glass for sale in the village," but he was disgruntled that it "was not more than a quarter as large as the page on which I am writing." Ohio native, and later governor of Wisconsin, Lucius Fairchild (1831–96) traveled to California in 1849. He wrote to his family from the city of Sacramento to let them know of his safe arrival. He reported that his traveling companions informed him that he had gotten taller during the journey, but Fairchild claimed that he did not know whether it was true or not because he had "no looking glass" in which to see himself.[21]

Men needed access to mirrors when they were away from home as well. The earliest American evidence of a "travel mirror" designed for men comes from the period of the French and Indian War (1754–63). A small glass mirror was

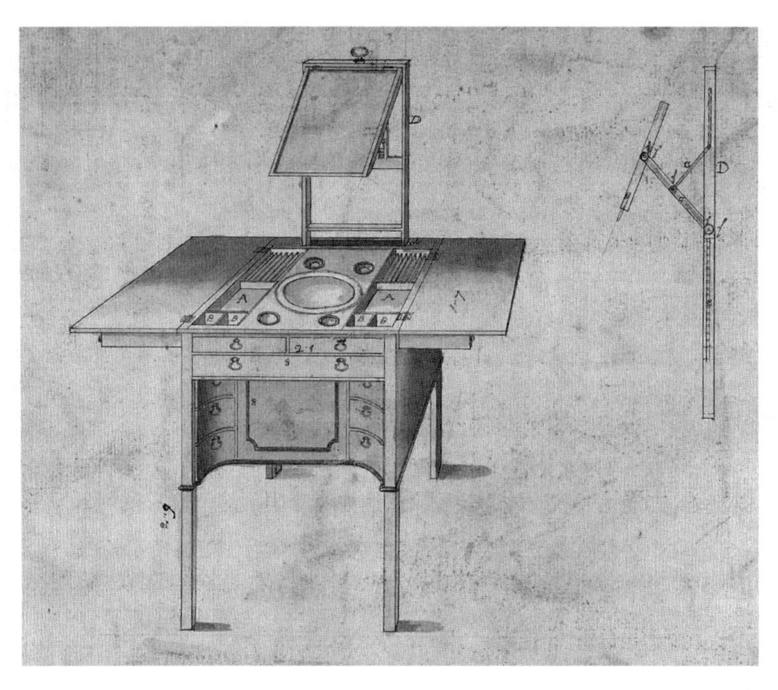

Figure 4.7. Shaving table. From *Chippendale Drawings*, vol. 2, ca.1753–54. Courtesy of the Metropolitan Museum of Art, www.metmuseum.org.

used as the base plug in a powder horn carried into that conflict. [22] The animal horn acted as a protective case for the glass, which could have been used for signaling or for grooming. A similar powder horn was used during the American Revolution and is dated 1777 (fig. 4.8). During the American Revolution and the War of 1812, men kept mirrors or shaving boxes (that held shaving supplies) with them in camp (fig. 4.9).[23]

By the time of the American Civil War, men commonly carried mirrors with them to their wartime duties (figs. 4.10, 4.11). According to Carlton McCarthy's 1882 book *Detailed Minutiae of Soldier Life in the Army of Northern Virginia, 1861–1865*, a looking glass was one of the standard items found in a Virginia soldier's knapsack.[24] Union soldier Henry Howe (1841–1900) wrote home to his mother and sister and thanked them for the "parting gift" they had given him of a looking glass before he had left for the war. "I have it now hanging in my room," Howe wrote, "and use it every morning." Union chaplain Hallock Armstrong (1823–1904) had only "a pocket mirror" to rely on while at camp but wrote to his family that it was "always at hand" and "well used."[25]

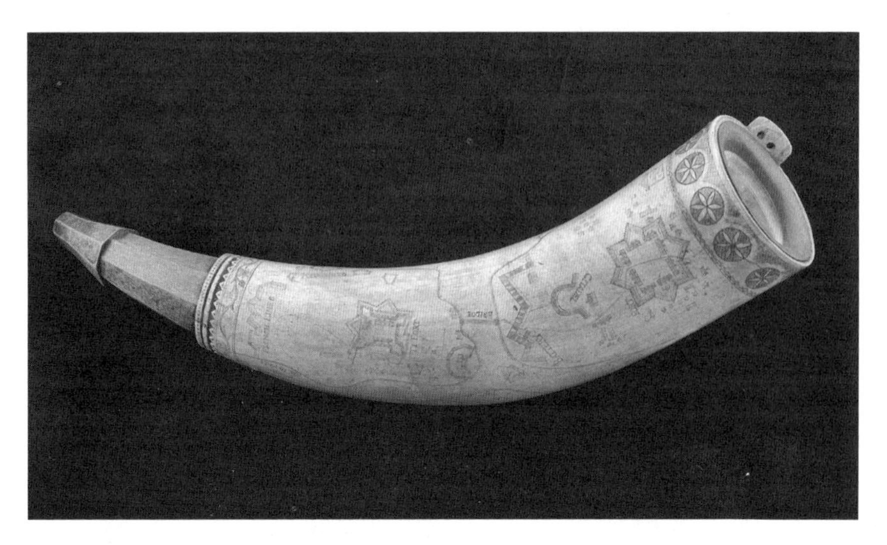

Figure 4.8. Powder horn with mirror stopper, no. 1922.003. Inscription reads: "WHAT I CONTAIN SHALL FREELY GO: TO BRING A HAUGHTY TYRANT LOW. / JOHN:CALFE. / HIS HORN, MADE AT / Mount Independance, Apl 1777." Courtesy of the New Hampshire Historical Society. Gift of Lizzie E. Crowther.

Figure 4.9. Shaving box, War of 1812, no. H14199. Courtesy of the Ohio History Connection.

Figure 4.10. Thomas Jefferson Brengle Civil War pocket mirror. Courtesy of the Stevens Memorial Museum, Salem, Indiana.

Not all men brought mirrors with them to the Civil War. Some expressed surprise when they discovered that one was not being provided for their use. When North Carolinian Louis Leon (b. 1841) entered the Confederate Army in April of 1861, he was unprepared for the spartan quarters that awaited him: "Lo, and behold! horse stables with straw for bedding is what we got. I know we all thought it a disgrace for us to sleep in such places with our fine uniforms—not even a washstand, or any place to hang our clothes on. They didn't even give us a looking-glass." Other men found living without a mirror added to the sense of isolation brought on by being away from the regular rhythms of life at home. Captain and Assistant Adjutant-General of the United States Volunteer Army William Thompson Lusk (1838–97) described his experiences at the Union headquarters of the seventy-ninth regiment in Virginia where the men were "quite shut off from . . . the outer world." Entertaining family, however, reminded Lusk that "there was still remaining communication with the world." He wrote that as a "result of the lesson" he "bought a looking-glass and combed the snarls out of my hair."[26]

While wartime might seem like a period during which appearance would matter less, men's desire to know what they looked like remained strong. Ohioan Osborn Oldroyd (1842–1930) was nineteen years old when he began his service in the Union army. He told the story of a fellow soldier, Calvin Waddell, who took "from his pocket a small mirror and was combing his hair and moustache"

Figure 4.11. Gen. Lawrence Bryan O'Branch Civil War mirror. Photo by the North Carolina Museum of History.

immediately before the fighting began at the Battle of Raymond (1863). One of his fellow soldiers teased Waddell for his behavior, saying, "You needn't fix up so nice to go into battle, for the rebs won't think any better of you for it." The men undoubtedly would have engaged in further merriment after the skirmish had Waddell not died in the fighting that day.[27]

By the mid-nineteenth century, white men had spent their lives around looking glasses. They used them in their homes and regularly carried them to war. They knew that mirrors accurately extended the limits of their vision by enabling them to see themselves. In daily life men used this ability to shape aspects of their personal front as they encountered their mirror selves. Yet white men did not interpret their encounters with their mirror selves in the same ways that women did. Under normal circumstances men did not convert what they observed in the mirror into an assessment of the acceptability of their mirror selves, nor did they judge the mirror self against an "objective" standard of attractiveness as women did. When men mentioned mirrors, they were much less likely than women to mention their mirror selves. Sometimes, however, extreme circumstances led men to assess their own reflections.

Illness or injury were situations in which white men commented on what they saw when looking in the mirror. In 1831 Sarah Judson (1803–45) and her husband were serving as missionaries in Burma. Sickened by an unnamed ailment, her husband looked in the mirror and "at once [saw] symptoms of his approaching dissolution." According to Judson he announced without emotion that he had "altered greatly" and was "sinking into the grave very fast—just on the verge." Charles Dabney, a young southerner from Pass Christian, Mississippi, returned home in 1853 after studying law at Harvard for two years, only to fall ill with yellow fever. Several members of his family gathered at his bedside. Dabney "asked his mother to hand a looking-glass to him." His mother "held it before his face, and he was shocked to see the blood on his mouth. At once he prepared himself for the death that could be but a few hours off." Wealthy antebellum South Carolina slaveholder and former governor James Henry Hammond (1807–64) was taken seriously ill in 1864 and was visited by his son and a doctor. Although for an hour they talked about "various topics," everyone was on edge. After the doctor left, Hammond confided in his son his concern, asking "what it all meant—the Doctor's sudden coming and mine, and Mother's anxiety?" Having been scrutinized for an hour by these worrying individuals, Hammond asked his son for a mirror so that he could see for himself: "He bade me raise and prop him in his bed, and hand him a small looking glass. He told me he did not feel as badly, that Mother's and the Doctor's anxiety was unnecessary and, looking at himself, remarked how well he looked, that he looked better than the day before, better perhaps than in twenty years past—his complexion clearer; but, on inspecting his tongue which was dry, engorged and darkened, 'ah,' said he, handing back the glass, with a depressed countenance, "that's a very bad tongue, very bad: lower me." After his son lowered him back in the bed, Hammond asked him to stay. It was now clear to Hammond that he was dying. Telling his son, "I may live three days; it may be only three hours," Hammond detailed where he wished to be buried.[28] Some men found that as the end of life approached, the mirror became a useful tool in judging how quickly death drew near or perhaps met their desire to see one last glimpse of themselves.

A young Civil War soldier whose face had been seriously injured by two bullets that had "passed through his cheek and jaw . . . , knocking out the teeth on both sides and cutting the tongue in half," also expressed serious concern about his mirror self. As Charlestonian Phoebe Pember, matron of the Chimborazo Hospital in Richmond, Virginia—who cared for him there—mused, the wounds were not fatal, but "fatal wounds are not always the most trying." After several days of care, the young man was able to speak clearly for the first time. His first request, Pember noted, "was that he wanted a looking-glass to see if his sweetheart would be willing to kiss him when she saw him." Regardless of what they may have actually thought, the nurses "all assured him that she would not be worthy of the name if she would not be delighted to do so."[29]

Not all shocking reflections were the result of sickness or injury. Sometimes the mirror self could be a source of unexpected mirth for white men. Naturalist John K. Townsend (1809–51) traveled west to the Columbia River in Oregon Country in 1834. After a long period of travel during which little time was spent being attentive to their appearance, Townsend and his companions drew near to a place where they knew they would encounter people "in whose eyes we wished to appear like fellow Christians." Preparing their public selves for this encounter included washing, dressing in fresh clothes, and shaving. After completing these tasks, Townsend surveyed himself in a mirror "with great self-complacence and satisfaction." Townsend "admired" his "appearance considerably" but noted his amusement at what shaving his beard had revealed. The skin beneath it was quite light while the rest of his face was dark: "the lower portion being fair, like a woman's, and the upper, brown and swarthy as an Indian." In a situation in which we can only imagine a white woman experiencing shame had she undergone a similar transformation, Townsend "could not refrain from laughing" at what the mirror revealed. Despite the fact that his face had taken on quite an unusual appearance, Townsend was not only amused but still pleased with his mirror self. While access to his mirror self provided Townsend with the critical information that he did now appear acceptably "Christian," his playful recounting of the unexpected discoloration of his face suggests a gendered understanding of his public self that was able to discount such an unusual appearance without fear of criticism.[30]

One final category of men's encounters with their mirror selves does not include men's thoughts about what they saw in their reflections, but it does reinforce what a key object looking glasses had become for white men. Some white men who committed suicide used mirrors to prepare to commit the act or in the commission itself. James Purrinton was a resident of Kennebec, Maine, who killed himself, his wife, and six of their eight children in 1806. A week before the murder-suicide, his daughter (who would survive the attack) observed him taking his butcher knife and sharpening it. He then "stood before the looking glass, and with his left hand seemed to be preparing his throat for the knife." Similarly, diarist and lawyer George Templeton Strong recorded an account of Nathaniel Prime's suicide in November of 1840. Prime "went to his room at two o'clock and appears to have taken up and read his prayer book, then went before the glass, cut his throat coolly and steadily from ear to ear, replaced the razor in its case, and then walked into the next room, and there fell." As late as 1902, the *New York Times* reported the case of John Fitzpatrick, "a prominent potter," who had been disfigured by smallpox and committed suicide as a result. The *Times* reported that "he stood for several minutes looking at himself in the mirror," immediately before taking a deadly dose of arsenic. For those contemplating—and eventually committing—suicide, perhaps encountering their mirror selves helped them imagine what they planned to do and find the determination to follow through. Whatever the reason, all of these men sought out their mirror selves and took a long hard look at themselves as they faced a moment of crisis.[31]

White men also found uses for mirrors beyond observing their own faces and bodies.[32] During the Civil War, soldiers developed several military uses for looking glasses, all of which underscored their belief in the reliability of the information mirrors provided. Published in 1893, *The Soldier in Our Civil War: A Pictorial History of the Conflict, 1861–1865*, included fifteen illustrated "Hints to Soldiers in the Camp and on Campaign." Among these illustrations one could find practical advice on the "Safe Mode of Sleeping with a Loaded Gun" and "How to Secure a Prisoner, " as well as an example of how a soldier might use a small looking glass for signaling to other soldiers (fig. 4.12).[33] Confederate soldier Samuel Watkins recalled how the "Yankees would hold up small looking-glasses, so that our strength and breastworks could be seen in the reflection of the glass." According to Watkins, these soldiers "also had small mirrors on the butts of their guns, so arranged that they could hight [*sic*] up the barrels of their guns by looking through these glasses, while they themselves would

Figure 4.12. "Signaling with a Piece of Looking-Glass." Paul F. Mottelay and T. Campbell-Copeland, eds., *The Soldier in Our Civil War: A Pictorial History of the Conflict, 1861–1865*, vol. 2 (New York: Stanley Bradley Publishing, 1893), 280.

not be exposed to our fire." Union soldiers attempted a more daring military application of looking glasses as well. Osborn Oldroyd recalled how his fellow combatants built a "tower, ten or twelve feet high" where a single man could sit and observe—using a looking glass hung "in such a position as to catch and reflect"—the "interior of the enemy's fort and rifle pits." Through such observation, Oldroyd noted, "every man and gun could be counted." Unlike other military applications of looking glasses that could be used from safer locations, this one required the lookout to be quite vulnerable. For that reason, the tower did not last long, being "too conspicuous and dangerous for use."[34]

White men also extended their belief in the mirror's accuracy of reflection to the realm of political and moral discourse by embracing the mirror as an appropriate metaphor to invoke truthfulness. In the same way that mirrors served as a tool for rational enlightenment about their own bodies, here the mirror became a metaphor for rational enlightenment of the mind. In these cases men made use of a metaphorical mirror to criticize someone who was often a political or moral opponent, to show that person the truth as perceived by the writer in the hopes of enacting a conversion. The mirror was to be the vehicle by which the person being criticized would reevaluate the situation and come to see, and believe in, the author's perspective. As early as 1676, a critique of Puritan New England's harsh treatment of dissenters was printed in London entitled "A Glass for the People of New-England." In this mirror (his writing), the author hoped Puritan New Englanders would "see themselves and Spirits, and if not too late, Repent and Turn from their Abominable Ways and Cursed Contrivances." "The Drunkard's Looking-Glass: or, A Short View of Their Present Shame, and Future Misery" (1728) sought to reform those who drank too much by showing them the error of their ways. Published in 1764, "A Little Looking-Glass for the Times; or, A Brief Remembrancer" sought to reform Pennsylvanians; the author intended that this work would "serve to give any such Persons a Glimpse of their own Faces; and thereby in anywise tend to raise in them a Curiosity for examining more strictly into their Conduct and present Manner of Life." Two weeks into the presidential election of 1800, John Adams wrote to his wife, Abigail, expressing his concern about the state of the nation if his opponent, Thomas Jefferson, were to defeat him. Adams wrote, "By that time this Nation if it has any Eyes, will see itself in a Glass. I hope it will not have reason to be too much disgusted with its own Countenance." Adams assumed that what people would see in this metaphorical mirror would be the truth of the situation they had enacted by electing Jefferson to the presidency. In 1807 "The Mirror of Misery; or, Tyranny Exposed" was published, a scathing critique of the slave trade and slavery. The tract challenged its readers: "Let now every honest man lay his hand on his breast, and seriously reflect," presumably in the titular "mirror of misery," about "whether he is justifiable in countenancing such barbarities; or whether he ought not to reject, with horror; the smallest participation in such infernal transactions." The truth about slavery was the mirror in which people would find the courage to challenge this hideous institution, the author argued. Because white men believed that the mirror reflected the truth about themselves when they sought out their mirror selves, they were able to expand on this belief by metaphorically using mirrors as fulcrums of truth in the world of ideas as well.[35]

Despite this wide range of both literal and figurative uses of the mirror, white men downplayed their dependence on looking glasses to the point of occasionally denying making use of this object at all, something no American woman I have encountered in doing this study ever asserted.[36] In 1844 the *Southern Quarterly Review* quoted an attorney and judge, Mr. Mansfield, who claimed, "I have not seen my face in a looking glass for thirty years." In 1856 Horace Greeley (1811–72), editor of the *New York Tribune*, wrote a letter to fellow journalist James Pike in which he asked Pike if he had ever woken up in a place where he was afraid to "look in a glass when you got up in the morning for fear of seeing a scoundrel?" Greeley observed that such a situation could never happen to him because "I never patronize looking-glasses."[37] The need these men felt to distance themselves from this fundamentally quotidian object suggests how feminized it had become in American culture. These men denied having a speck of vanity in themselves by claiming to have no interest in their own appearance or in the object most associated with acquiring knowledge about the physical self, the mirror.

Considered together, white men's and women's wide range of mirror usages revealed their trust that mirrors reflected back an accurate representation of what was placed before them. In enabling them to gain visual access to important information about themselves and their world, mirrors increased their reliance on and trust in what they could see. Although both men and women regularly used mirrors in the creation of their public selves, how they did so revealed and exacerbated the ways in which patriarchal gender roles shaped their experiences. For white men the mirror was yet another tool that enabled them to increase their mastery over themselves and their world. Although the mirror gave white women better mastery over the personal front, many women seem to have been ultimately mastered by the mirror's gaze as they sought to monitor their public selves. In this way the looking glass enabled women to see themselves through men's eyes—to employ the male gaze—as they judged the acceptability of their mirror selves. This is not to say that women did not use mirrors to increase their self-knowledge or that women only ever employed the male gaze when they looked into mirrors. But this element of women's mirror usage was distinct from how men interacted with the looking glass. For both men and women, however, the mirror was a critical tool in creating white gender identities, both reflecting and shaping patriarchal gender roles.

Troubling Visions

In their reliance on mirrors in daily life, white men and women trusted what their eyes revealed when they encountered their mirror selves. But as we saw in chapter 2, mirrors contained within them a history of having been associated with magical beliefs and practices that challenged the idea that the mirror could be trusted to be an accurate reflector of the material world. The proliferation of accurately reflective glass mirrors in early America made an object that had often been a diviner's tool into a common personal possession and piece of household furniture. Beliefs about reflective powers did not disappear as a result of this material transformation of the mirror. In some cases the proliferation of mirrors seems to have actually strengthened putatively premodern beliefs about reflection and created a need for new practices that could contain the power this object continued to possess. Moreover, the proliferation of looking glasses gave many more people access to an object through which these beliefs could be practiced. Although these beliefs did not disappear, evidence about the practices is quite scarce because they were usually hidden from public view.[38] But some evidence involving mirrors does remain.

One power attributed to reflective glass was the ability to divine the future, specifically to identify one's future mate. The earliest evidence of this practice in North America comes from Puritan New England, where the minister John Hale, from Beverly, Massachusetts, recounted in a 1702 publication the story of a young girl's visit to a "Doctor" who claimed he "would shew Maids their future Husbands in a glass." The young girl—now a married woman—believed he had shown her the image of the man who became her husband.[39] This practice either persisted or was revived at some point, because a century later, New Jersey native Lydia Sexton (1799–1872), licensed as a United Brethren preacher in 1851, recalled a "game" she had played during her childhood, at the home of her friend Nelly Conselus. Sexton described how "Nelly proposed that we should try our fortunes in some way before going to bed." One of the suggested methods for seeing into the future was "to look down the well, some one holding a looking-glass, face downward, over your head; you would see your intended in the glass, reflected in the water." By the time she told about this practice, as an adult, Sexton strenuously rejected divination, distancing herself from these practices by insisting that she "had no faith in the 'black art,' and only did it for amusement." To emphasize her doubt in the efficacy of these practices, Sexton continued: "the imagination and wish, of course, had much to do with it; and sometimes this ideal would be fancied. But the whole thing is a very injurious and dangerous superstition, especially among young persons of that age, so susceptible of wild, fantastic, and superstitious impressions."[40] Despite her protestations, Sexton's story revealed that the practice still existed around the start of the nineteenth century.

An 1844 edition of the *Southern Literary Messenger* included a very similar fictional account of a group of young girls who gathered to "try their fortunes" before a looking glass. They took the glass and held it over a local well; one of their mothers had done this as a girl and had seen her future husband. Each girl took the glass in turn, some of them returning "with a gleam of satisfaction, while others looked indifferent, or disappointed." Finally, the eldest girl in the group, Julia, grasped the glass and "looked on its polished surface with great intensity." Before long, Julia became "deadly pale." She would tell no one what she had seen. Later she confided in her sister that she had seen "an open grave." Not long after this, according to the story, the prophecy was fulfilled: young Julia "closed her eyes and slept forever."[41] As late as 1867, an advertisement in *Harper's Weekly* offered mirrors especially for divination: "A MODERN MIRACLE.—The magic Mirror, by which either lady or gentleman can produce their future husband or bride, the greatest thing yet out. Two mirrors sent to any address free by mail. Enclose 50 cents to DRAKE & CO., 38 Broadway, N.Y."[42] This "magic Mirror" was labeled as "MODERN," but in reality, it linked its viewers back to the long tradition of using reflective surfaces for divination.

Another practice that was rooted in the belief that mirrors possessed power was turning mirrors around to face the wall or covering them with a cloth at the time of a death.[43] In his classic early twentieth-century work *The Golden Bough: A Study in Magic and Religion*, the Scottish anthropologist James Frazer linked this practice in Europe to the fear "that the soul, projected out of the person in the shape of his reflection in the mirror, may be carried off by the ghost of the departed, which is commonly supposed to linger about the house till the burial." Frazer found that this was a common custom around the world, including in Germany, Belgium, England, and Scotland. This practice likely flourished only after glass mirrors became permanent items of furniture in households. Only on the wall would mirrors need to be covered with a cloth or turned around to shield potential observers from dangerous reflections.[44] Prudence Punderson's late eighteenth-century embroidery is the only visual evidence extant of the practice of covering mirrors at the time of death during this early period. She depicted a white woman in the three stages of life: as a baby, watched over by a woman of African descent; as an adult woman at a writing table; and, finally, in death. Behind

Figure 4.13. The First, Second, and Last Scene of Mortality, embroidery, by Prudence Punderson. Gift of Newton C. Brainard, Connecticut Historical Society, no. 1962.28.4.

the coffin, a mirror—very similar to one owned by Punderson's family—is shrouded in a white cloth (fig. 4.13).[45]

In the nineteenth century the White House followed this practice. At the time of President William Henry Harrison's death in 1841, the mirrors in the East Room were covered: "The splendid mirrors, which rose almost to the lofty ceiling, reflecting on every side the brilliant crowds which often thronged this room, now refused to look upon the scene before them, and buried their polished bosoms in the habiliments of sadness." After President Lincoln's death, on April 15, 1865, his body lay in state in the White House before the funeral train took him to Springfield, Illinois. The extensive preparation to the East Room of the White House included that "the windows at either end of the room were draped with black barege, the frames of the mirrors between the windows, as well as those over the marble mantles, being heavily draped with the same material. The heavy gildings of the frames were entirely enshrouded, while the plates of the mirrors were covered with white crape."[46]

Two individuals also recalled mirrors that were covered at the time of death in more ordinary white homes. Aunt Clussey, a formerly enslaved woman of African descent from Alabama, recounted how, when her master died in 1865, the mirrors were "drape[d]" in his house. Clussey explained that the mirrors and pictures had to be covered because if the deceased's spirit saw itself there would be another death in the house.[47] Born in 1877, Lloyd Douglas, a white man, recalled the practice from his rural Indiana childhood: "When there was a death in the family it was a country custom for some helpful neighbor to stop your clocks, probably as a courtesy to the deceased." Furthermore, Douglas recalled, someone would "turn all your mirrors to face the wall." While it was unlikely that Douglas had continued the practice—he couldn't remember why it had been done—his memory suggested that it was still done very late into the nineteenth century.[48]

Another superstition associated with mirrors was that breaking one brought bad luck.[49] The proliferation of larger glass mirrors—and the practice of hanging them from walls—would have made breaking a mirror much more common and may have contributed to the persistence of this belief, with which we are still familiar today. Writing in 1828, James Hall, a Philadelphia lawyer, writer, and newspaper editor who moved to the Ohio Valley after serving in the War of 1812, assumed that most American youngsters had been "advised of the danger of fracturing looking-glasses." Hall also asserted, however, that the fear of bad luck from a broken looking glass was childish, and he claimed that it would have been discounted by all adults. Hall went so far as to proclaim his belief that the American people were so far removed from such beliefs as to have "no national, and but few local superstitions." Hall argued that "the general diffusion of intelligence, has left no rank of society in absolute ignorance; and we find few individuals in that state of intellectual degradation which pervades the lower orders throughout a large portion of the globe." He believed that "almost all" Americans had "arrived at that degree of refinement, at least, which awakens the mind to inquiry, and brings it within the reach and influence of knowledge." In this state of refinement, Americans had set aside, Hall believed, such superstitious beliefs and practices.[50]

Hall wrote definitively that Americans had denounced superstitious beliefs, but other evidence suggests he may have overstated their marginalization. An 1858 column in *Harper's Weekly* explored a variety of beliefs associated with breaking a looking glass: "To break a looking glass is accounted a very unlucky accident. Should it be a valuable one this is literally true, which is not always the case in similar superstitions." This 1858 article linked the association between mirrors and bad luck with "an ancient kind of divination by the looking glass" and the use of mirrors "by magicians in their diabolical operations." Beyond simple "bad luck," breaking a looking glass was also associated with other misfortunes: "its owner will lose his best friend," and "breaking a looking-glass betokens a mortality in the family."[51]

Native Virginian and author Constance Cary Harrison (1843–1920) recalled in her memoir, *Recollections Grave and Gay*, an occasion when a broken mirror was linked to just such a tragic outcome. Harrison's cousin Hetty Cary and Brigadier General John Pegram were married in January of 1864. He returned, immediately after the wedding, to his headquarters near Petersburg, where Pegram served during the Civil War. Pegram died there in combat only three weeks later. His body was returned to the church where he and Cary had been married. Harrison recalled how, two days before they were married, Cary had come to her room to try on her bridal veil so that Harrison and their mothers "might be the first to see it tried on her lovely crown of auburn hair." After affixing it atop her head, Cary turned around to face the onlookers. At that very moment, the mirror into which she had just been looking "fell and was broken to small fragments." Although Harrison distanced herself from the belief that there was a connection between the broken looking glass and Pegram's death, she acknowledged that it was "spoken of by the superstitious as one of a strange series of ominous happenings."[52]

Again in 1858, another article in *Harper's Weekly* maintained that there "are grave people who are really seriously troubled if they tip over the salt—if they break a mirror—if they pass under a ladder—and a hundred other things that somebody must be continually doing somewhere." The author acknowledged that people wished they did not hold these beliefs, but they could not entirely rid themselves of them. The author noted, "It is a kind of superstition which is very prevalent, but is hushed up, because, although every body hasn't sense enough not to share it, every body has sense enough to be ashamed of it."[53] The person who recorded Lincoln's unsettling encounter with his mirror self that opened this chapter went even further and claimed that everyone did share a belief in these superstitions—that "flavor of superstition which hangs about every man's composition"—as he described it.[54]

Although an undercurrent of denial ran through these accounts, they also revealed that some white Americans did continue to hold beliefs in the power of reflection.[55] By the mid-nineteenth century, however, white Americans began to comment on another kind of visual uncertainty that their mirrors produced that discounting beliefs in "superstitions" could not eradicate. This kind of visual uncertainty occurred when men and women saw something unexpected in their encounters with their own mirror selves. Although nineteenth-century men and women trusted mirrors to provide an accurate representation of themselves, there were times when men and women experienced a profound disjointedness between what they thought they looked like and what the mirror showed them to be.

Nineteenth-century white Americans were the inheritors of a long tradition that put great confidence in human vision more generally as what had been understood for centuries to be the "noblest sense," the one that offered the most accurate and truthful access to information about the world.[56] By the nineteenth century, however, pressing questions about vision challenged its primacy. Rooted in increased scientific understanding of the physiology of the eyes, occurrences like retinal afterimages—shapes and colors that can be "seen" by a closed eye—transformed the body itself into a site of the production of vision that might have no reference point outside of that body. Could vision really provide the most reliable representation of the world if human eyes saw things that did not exist outside the body?

This nineteenth-century understanding of vision as being fundamentally "embodied"—that is, situated within and subject to the imperfections of the human body—raised serious questions about its verisimilitude. One popular technology that laid bare the unreliability of the eyes was the stereoscope. Developed in the 1830s, alongside the emergence of daguerreotypy, this device allowed an observer to hold at a fixed distance two pictures of the same scene taken from slightly different perspectives that yielded, from the viewer's perspective, one, three-dimensional image (fig. 4.14). Stereoscope users knew that they were not seeing the scene itself but a representation of it made to appear real by the working of their own eyes. Stereoscopes revealed that human vision could be fooled—in this case, into thinking that a two-dimensional picture was actually a three-dimensional reality.[57]

Mirrors had taken up a prominent place of residence in most white American homes by the time that vision came under these challenges in the nineteenth century. While mirrors had certainly contributed to the visual confidence of the seventeenth and eighteenth centuries, they were also sites at which faulty, embodied vision could be observed and monitored. As Ralph Waldo Emerson's daughter Ellen Tucker Emerson (1839–1909) observed in 1853, it was the

Figure 4.14. Holmes Stereoscope and stereograph card. North Carolina Collection, Wilson Special Collections Library, University of North Carolina at Chapel Hill.

mirror that enabled her to assess the status of her worsening vision. In a letter to her sister, Emerson wrote, "My eyes are beginning to trouble me. For four or five weeks when I have read or studied or written by lamplight my eyes would grow stiff in a minute or two . . . when I looked at them in the glass I found they were red."[58]

Antebellum white Americans were also experiencing a rapidly proliferating visual culture that strained vision and made it more likely to fail, something Emerson may well have been experiencing when her eyes grew "stiff" as she read, studied, and wrote. Especially in urban environments, Americans faced new challenges to their vision from the dizzying array of print sources (e.g., newspapers and books, signboards, broadsides), as well as from situations and activities (e.g., overstimulation and environmental factors) deemed deleterious to ocular health. Spectacles and other corrective practices became more important during this period as white men and women scrambled to find ways to address visual weaknesses. White Americans questioned the belief that vision was the "only valid way of knowing the world" as they faced increasing demands placed upon their eyesight.[59]

When concerns arose in the nineteenth century about whether human vision really was the most accurate way to know the world, devices like the stereoscope, which emphasized the fallibility of human vision, fell out of fashion.[60] But the mirror had become too ingrained a part of daily behavior to be so easily discarded. White men and women's daily dependence on the mirror as a tool that supplied critical knowledge about the self meant that they could not completely disregard questions about the veracity of vision that sometimes met them in the mirror's gaze.

Questions about whether one's mirror self produced an accurate representation of one's actual self could arise for a variety of reasons. A physical or emotional transformation might reveal itself in the mirror, creating a gap between what observers thought they looked like and how they actually appeared. Louisa May Alcott fell seriously ill with typhoid fever in 1863. When she was well enough to look in a mirror, she found a "queer, thin, big-eyed face" quite unlike her usual mirror self. Alcott said that she "didn't know myself at all."[61] Maria Lydig Daly (1824–94) encountered an unexpected mirror self during an unusual visit to a friend's house in 1861. Daly, born into a prominent New York City family, married a judge in 1856. In December 1861, with her husband away in Washington, DC, Daly made an overnight visit to Staten Island to see a friend. For reasons she does not divulge in her diary, Daly felt "tired" and "melancholy" during the short trip and noted that the "people were all strange" on the island. Daly's cheerlessness was exacerbated by the mirror in her friend's home. The mirror was, she reported, so "bad" that when Daly looked at her reflection in it, she "looked so old and ugly" as to be "distressed." Returning home—and to her own mirror—lifted Daly's spirits considerably, for she found that "I am just the same person I thought myself before I left home."[62]

The experience of war could cause a similar disjunction between one's internal sense of self and the mirror self. Some Civil War soldiers encountered strangers when they looked in the mirror as they experienced harrowing circumstances that changed their exterior appearance even as they held on to their interior sense of self. Daniel Hundley was a Confederate soldier who attempted an escape from the Johnson's Island, Ohio, prison camp. After making it off the island, Hundley spent one night at a hotel. After the night clerk had shown him to his room, Hundley "took the lamp and looked at myself in the mirror. The face reflected back to me was as strange as if I had never seen it before. My most intimate friends would never have recognized me. It was a bloodless face, white as a ghost, with great, staring eyes, from the hollow depths of which seemed to burn a wasted flame like the last flickerings of a midnight taper."[63] In 1862 the Reverend James K. Hosmer joined a Union regiment that would be active in Louisiana over the next two years. Entering a home there that had been ransacked by Union forces, Hosmer confronted himself in a "large mirror." He "caught sight of a very swarthy and travel-stained warrior, whom I should never have recognized." Hosmer was a minister, and his self-identity as such did not conform to the reflection of the "warrior" he saw in the glass. He acknowledged that if he had not known for sure he was looking at himself, he would "never have recognized" the man in the mirror. He "hurried out with an uncomfortable feeling."[64] If someone who identified as a minister could meet a warrior in the mirror, how else might Hosmer's eyes lie to him in daily life? No wonder the moment unnerved him.

While the previous accounts have recorded moments when a person saw something unexpected in the mirror, sometimes what surprised nineteenth-century Americans was that they looked exactly the same as they always had in the looking glass. Sarah Morgan recorded one such experience in her diary in 1863. Morgan found herself in an uncomfortable social situation with several other women at the end of which she made promises she did not intend to keep about being "sociable" and made false "professions of esteem." She worried that the other women had been able to see through her public self to her interior feelings. After the visit, she consulted a mirror to see whether how she felt was visible—whether she looked like she had been "split in two like a young spring chicken, and deprived of heart and brains." Despite the inner turmoil, the mirror self seemed to have shown no sign of change; Morgan concluded, "I suppose it was all right."[65]

These disjunctures had societal implications in the increasingly mobile setting of mid-nineteenth-century America. As Americans began to encounter more frequently people with whom they had no preestablished connections, they increasingly doubted whether they could trust what their eyes told them about these strangers. Intimate knowledge about how their own mirror selves could hide their interior thoughts and feelings—and that other people could be hiding their true thoughts and feelings behind a similarly false public self—undoubtedly contributed to this uncertainty. White Americans had increasing opportunities for mobility by 1850. One study of men in the United States found, for example, that "20% of rural males moved to urban places over the 1850s."[66] American cities grew during this period as many moved into urban areas both from other parts of the nation and from abroad. Urban areas were notoriously anonymous as new immigrants to the United States, citizens from more rural areas moving to the city, and visitors jostled one another on busy city streets. As historian Karen Halttunen has observed, the American city "presented a serious problem: how could one identify strangers without access to biographical information about them, when only immediate visual information was available?"[67] As white Americans became more mobile, people increasingly lost the ability to know the family history and life stories of the people with whom they did business or socialized. Because, as Halttunen describes it, "people require information about those they meet, in order to avoid both psychological and physical damage," concerns heightened after 1850 about whether "visible outward signs" could reveal "inner moral qualities." These anxieties manifested themselves in, and were in part prompted by, encounters with the mirror self.

The dislocations caused by the Civil War also contributed to the problem of anonymity as people traveled vast distances and fought against (and sometimes with) those about whom they had no previous knowledge. As historian Stephen Berry has argued, southern Civil War soldiers resisted the loss of their sense of self even as they were caught up in a wartime machine that functioned best when each man was little more than an "interchangeable part of a larger unit." This loss of self "seemed a part of the air they breathed or the dirt they slept in, elemental, an inalienable part of the project itself." In this context the mirror confirmed the "depersoning" of soldiers. Confederate soldier James Williams complained that he had "no opportunity of judging of my appearance." Even though he had seen "the reflection of a dirty dust begrimmed [*sic*] face once or twice in a glass," he seemed disconnected from his mirror self.[68] Virginian Jane Woolsey had a similar experience on the home front during the war. She remarked how "when we came to town last week the streets seemed full of anxious and haggard faces of women, and when I caught sight of my own face in a shop glass I thought it looked like all the rest. The times are not exactly sad, but a little oppressive."[69] If everyone looked the same, then using visual clues to determine the character of the people one met was rendered impossible.

The moral of "My First Patient," a story published in *Harper's Weekly* (1865), suggested how far removed white Americans were from Enlightenment-era confidence that looking glasses increased their visual mastery of the world by the time the Civil War came to an end. The story was about a young doctor who was seeing his very first patient. The doctor was trying to catch the wife of his patient, who he thought was poisoning her husband. He had read "in a French book once how a great criminal was detected in the act of murdering his master by a mirror—that the person watching had seen the reflection of the assassin stifling the cries of his victim with a pillow." A readily available mirror—which was, by the time of the story's publication, an almost ubiquitous household item men and women relied on daily—was placed in the man's bedroom in such a way that the doctor would be able to see the wife if she entered the bedroom to administer a dose of arsenic. As he suspected, late in the night, the doctor observed the young woman entering the room and giving her husband something to drink—which the doctor assumed was the poison. As it turned out, however, the young doctor was wrong: the wife was not trying to poison her husband. The story concluded with this admonition: "Don't take what you see in a looking-glass to be a fact, or more than partial evidence of the truth and worth of any thing."[70]

An earlier generation of white Americans would have disagreed with the moral of this story. As mirrors had become common household items over the course of the eighteenth and nineteenth centuries, white men and women had embraced them as reflective devices that produced an accurate image of men and women's public selves, although earlier beliefs in the power of mirrors did not completely disappear. By the mid-nineteenth century, however, as white Americans faced increasing questions about how much they really could trust vision to reveal to them accurate and meaningful information about the world, the mirror became an important site at which the limits of embodied vision were experienced. Mirrors revealed how much the eyes could be fooled when the one thing that should have been most recognizable to a person in the mirror—his or her mirror self—appeared as a stranger. Discomfort with reminders of the limits of embodied vision might have prompted white men and women to discard this object, but it had become too entrenched in their daily lives to excise. As white men and women increasingly encountered actual strangers, especially on the streets of the rapidly growing American city or on a battlefield far from home, knowing that one could be surprised or even fooled by the appearance of one's own mirror self could only intensify concerns about what others might be concealing behind their public selves as well.

CHAPTER FIVE

Fashioning Whiteness

By the late eighteenth century, Europeans and their descendants in America had created a common racial identity—whiteness—that elevated them alone as citizens in the new nation, to the exclusion of the peoples they categorized as others: Native Americans and African Americans. As literature scholar Valerie Babb has argued in *Whiteness Visible: The Meaning of Whiteness in American Literature and Culture*, European Americans found that the best way to "vindicate their claim to the New World" was by "conceiving of themselves as a chosen white people." Whites' creation of these racial categories can be seen clearly by the mid-eighteenth century, when people of European descent across North America commonly referred to themselves as "white" to distinguish themselves from those who were "red" or "black."[1] Europeans built these racial identities in the New World with a variety of strategies across time and place. While much of this process unfolded in the realm of politics, law, and the economy, the more commonplace elements of life also contributed.

Objects of the material world helped to build and solidify the American racial hierarchy. Describing whiteness as "at most, a piecemeal wall built from everyday things," literature scholar Bridget Heneghan contends that, in the period from the Revolutionary War to the Civil War, European Americans increasingly identified as white and preferred white objects—ceramics, gravestones, women's clothing, and houses—because these things solidified their racial identity. Whites also used darker objects, Heneghan argues, to mark nonwhite racial identities. While some enslaved people, for example, "may have used discarded, unmatched dishes from the planter's household, they were more often issued dark, undecorated earthenwares." In these ways, whites used light and dark objects to manifest the racial boundaries they had created in the material world.[2]

Although mirrors fall outside of Heneghan's categories of light and dark objects, it is certainly true that wealthy white Americans possessed larger and finer mirrors than anyone else in North America, and the difference between one of these elaborately framed and ornate looking glasses and a ragged shard of looking glass was significant and could be used to mark difference. But because mirrors played such a core role in shaping individual identity, regardless of their size or decoration, they helped to fashion whiteness in other ways. Although, as we have seen, whites actually had a much more complicated relationship with reflection than they often acknowledged, in seeking to mark racial differences using mirrors, whites emphasized their own mirror use as providing accurate information about themselves and their world to aid in establishing their mastery over these realms. The emphasis nineteenth-century whites placed on linking mirror use to truth and visual mastery over the world helps explain why they denied the more troubling visions they encountered in mirrors. Whites believed that their own use of the mirror was the "correct" or "natural" one and thus perceived any use that deviated from their established norm as a sign of inferiority. Whites sought to differentiate what they claimed to be their use of the mirror as a rational tool of self-discovery and extender of human vision from the use of the mirror by Native Americans and African Americans.

From the earliest era of their colonization of North America, whites argued that Native peoples were less civilized than peoples of European descent because of their "unbalanced gender roles." Native women provided a significant majority of the calories consumed through their labor, built and took responsibility for maintaining their homes, and generally seemed to European observers to work harder than the men. This represented an inversion to Europeans of how they had constructed gender identities in their own cultures.[3] As part of the larger argument that Native peoples in North America were less civilized because they did not understand "proper" gender roles, whites came to strongly associate Native American men with the mirror. In white society looking glasses were rhetorically linked to the vanity of women. Emphasizing Native American men's use of looking glasses fundamentally called into question their masculinity by linking them to what was, in the white view, a feminized object. European Americans frequently noted Native American men's public use and display of looking glasses. As we have seen, although white men used mirrors regularly, particularly in private domestic spaces, they downplayed concern about their appearance and minimized evidence of their mirror use.

White observers reported that Native American men felt no such constraint, commonly accusing them of excessive vanity and near obsession with the looking glass, in part because Native American men used mirrors where they could be seen by others.

The emphasis on Native American men's mirror use emerged as early as 1650, when a Jesuit missionary to the Hurons in New France noted Native men's use of mirrors. Father Bressani observed that because among the Huron, only men practiced painting, "it is the custom of men, and not of the women, to wear even in war little mirrors about their necks, or in the small pouches in which they carry the Tobacco." Bressani seemed to expect his readers to assume that Huron women would be the primary users of mirrors—based on European assumptions about mirror use—so he clearly explained that it was the "custom of the men, and not of the women," to carry mirrors with them at all times. Whereas Bressani commented on men's practice of keeping mirrors close by, French explorer and trader Nicolas Perrot emphasized how men used them. Perrot traded and explored in New France in the latter half of the seventeenth century and worked with the Jesuits. Historian Stewart Rafert considers Perrot to have been an "expert on day-to-day Indian relations."[4] Perrot declared that if Native men "had a mirror before their eyes they would change their appearance every quarter of an hour."[5]

Father Bressani was correct in his assertion that only Huron men painted, so we have little reason to doubt that he was also correct that the men carried mirrors with them. Given that among the Huron, women did not practice face painting, perhaps they did use mirrors with less frequency than their male counterparts.[6] But neither Bressani nor Perrot addressed the question of whether women made use of men's mirrors, had their own, or eschewed mirror use entirely. Rather, what came across most clearly in these accounts is that mirrors were *associated* with men. This association emphasized the difference between Native peoples and Europeans, among whom, even though both sexes made use of mirrors, the looking glass was strongly linked to women and their putative vanity.

The association of men with mirrors in Native societies across North America would continue throughout the eighteenth and nineteenth centuries. French Jesuit Pierre François Xavier de Charlevoix, who was in North America in the early eighteenth century, commented on his frustration with a young Native American man's willingness to let his concern for his appearance disrupt his work. This young man took him by canoe across Lake St. Claire, located just east of present-day Detroit, but did not allow the task of rowing to divert his attention from his appearance, according to Charlevoix. "At every three strokes of his oar" this young man "took up his looking glass to see whether the motion of his arms had discomposed the œconomy of his dress, or whether the sweat had not changed the disposition of the red and other colours with which he had daubed his face." The extent to which Charlevoix exaggerated remains unknown, but the impression that white readers of his account (first published in 1761) clearly got was of a Native man unnaturally concerned with his appearance.[7]

Swedish botanist Peter Kalm traveled in North America from 1748 to 1751. His writings from the journey were translated and published in English in 1770. Kalm described the "chief goods" of French traders at Montreal. He included looking glasses, noting that "Indians like these very much and use them chiefly when they wish to paint themselves. The men constantly carry their looking glasses with them on all their journeys; but the women do not. The men, upon the whole, are more fond of dressing than the women." Like Bressani a century earlier, Kalm made clear to his readers that it was the men who should be associated with looking glasses. Although Kalm specifically mentioned women, he only revealed that they did not carry looking glasses with them. He did not address what uses Native women actually might have made of these objects.[8]

The pattern continued in the nineteenth century. Trader Josiah Gregg's *Commerce of the Prairies* was based on his 1841 expedition into Native American territory around Santa Fe and was first published in 1844. His description in the "Indians of the Prairies" included a section on body adornments. He observed that, "with respect to dress and other ornaments," nothing less than the "order of the civilized world is reversed among the Indians." Women wore less paint and far fewer ornaments Gregg contended; it was the men who would spend "as much time at his toilet as a French belle, in the adjustment of his ornaments—his paint, trinkets, beads and other gewgaws." Gregg pushed further, claiming that for an Indian man, the "mirror is his idol: no warrior is equipped without this indispensable toilet companion, which he very frequently consults."[9] As Alexander Ross (1783–1856), a fur trader in the Oregon and Rocky Mountains, described it in an 1855 publication, young Native American men could be seen "all day in groups, with a paper looking glass in one hand, and a paint brush in the other. Half their time is spent at their toilet, or sauntering about our establishment. In their own estimation they are the greatest men in the world."[10] Gregg and Ross ridiculed and condemned Native American men's mirror use and face-painting practices and feminized them in white eyes.

White observers also noted what they considered an overestimation of the mirror's value by Native men. One such manifestation of the high value they ascribed to mirrors was the practice of reframing European mirrors. In addition to his mocking comments about how Native men used mirrors, Gregg chastised a male Native consumer who "usually took" a European mirror out of its original case in order to put it "in a large fancifully carved frame of wood, which is always carried about him."[11] Investing time and materials to frame a mirror that was already framed when they bought it seemed foolish to white men. Native men further revealed this perceived overestimation of the mirror's value in what they were willing to trade in order to acquire one. James Hobbs (b. 1819), who had spent time as a captive of the Comanche, described how nineteenth-century Comanche men would "give a horse for a piece of mirror." Moreover, Hobbs noted that if Comanches pillaged the house of a white man who had in his possession a mirror and a silver watch, they would value the mirror far above the watch. Hobbs described this imagined white man as an "unlucky trapper or emigrant," taking care to note that he had the mirror "for shaving"—a purpose Hobbs viewed as practical and therefore appropriate. The Comanche men, in contrast, would prize the mirror "as much the most valuable," according to Hobbs, for its "wonderful reflecting properties." Moreover, the Native American men appreciated neither the value of the silver watch nor of time consciousness, according to Hobbs, who believed that the watch would be "broken up, and the pieces made into nose or ear ornaments for the squaws and papooses." Hobbs juxtaposed the mirror desired by the Comanche men (of high value to white women) and the watch broken up and given to the Comanche women (of high value to white men) to underscore the reversal of gender roles he perceived among the Comanche and their lack of understanding what was to be valued in the material world.[12] In the 1890s one Blackfeet man lost his life because he rode back to retrieve his looking glass that he had left behind when his people had moved their village. Headed back to the previous village site, the man was ambushed and killed by a group of Shoshones. White observers would have perceived this as an extreme example of Native men's overreliance on the mirror.[13]

White association of Native American men with the looking glass also took visual form in photography. Late nineteenth- and early twentieth-century photographs of Native American men provide stark visual evidence of the ways in which whites had, for centuries, associated Native American men with mirrors. White photographers who had Native American men sit for studio portraits frequently posed them holding a mirror or wearing clothing decorated with mirrors. The photographs were often staged by the photographers, who "asked one Indian after another to pose in similar ways, hold identical objects, and even wear clothing not necessarily of the appropriate tribe." These photographs tell us more about what whites believed about Native Americans than they do about the Native people who posed for them.[14]

Sometime between 1884 and 1885, a Shoshone man named Moragootch had a studio portrait made. In the photograph he wears a vest decorated with metal beads, arm- and wristbands, and necklaces. His right hand cradles a mirror in a bootjack-style mirror board also decorated with metal tacks. From the perspective of white observers, the heavily modified looking glass suggested both his overestimation of the value of this item and his feminine concern with appearance (fig. 5.1). When Crow Yellow Dog posed for a portrait with an unidentified woman, he held a cased rifle in his right hand and a mirror, framed in wood with metal tacks and equipped with a leather handle, in his left (fig. 5.2). In a picture that included a woman, having the man hold the mirror sent an especially clear message to white viewers.

Whether women were present or not, several Native American men were posed holding both a mirror and a gun. In addition to Yellow Dog (Crow), both Gorosimp (Shoshone) and two unidentified Arikara men who posed together were so portrayed (figs. 5.3, 5.4). Perhaps white photographers intended the mirror and gun to work together to send a unified message. If the gun emphasized to whites the real threat that Native Americans had posed to white ambition as the United States moved west earlier in the century, the mirror confirmed white assumptions about Native peoples' otherness and their inferiority. Together these objects reminded white viewers of the appropriateness of removing people they believed to be inferior to them, who had, at one time, posed a threat to white American settlement in North America.[15]

Native American men might also be portrayed in photographs wearing clothing adorned with the circular paper mirrors that had for so long been a popular trade item. Spotted Elk and Little Horse, both of the Sioux nation, were photographed with a row of these small glass mirrors prominently displayed in the picture (figs. 5.5, 5.6). A photograph, later distributed as a postcard, by Byron Harmon (1876–1942) of Alberta, Canada, shows a Native man identified

Figure 5.1. Moragootch (Shoshone), between 1884 and 1885(?). Denver Public Library, Western History Collection, no. 32267.

only as "Bearskin Indian," wearing seven unusually large circular mirrors on his clothing (fig. 5.7). As is true in all of these photographs, it is impossible to connect these objects with the photograph's subject. Given how popular these round mirrors were in the Native American trade, it is quite possible that the photographers kept stoles adorned with these paper mirrors—like the ones featured in all three of these photographs—on hand for use by their photographic subjects. Adorning Native men with these mirrors in photographs reinforced the link between Native men and the feminized vanity embodied by the looking glass. It also reminded the white audience that Native people used

Figure 5.2. Yellow Dog (left) and unidentified woman (Crow), 1883. Photograph by Frank Jay Haynes (1853–1921). Courtesy of North Dakota State University Archives.

these mirrors in a way that whites did not—as adornment to clothing and other textiles. Whites could interpret this display as Native people misunderstanding how this object was "supposed" to be used, additional evidence to augment the spurious claim of Native American inferiority.

The public portrayal of Native peoples in white society definitively associated Native men with mirrors as early as the seventeenth century and continued to do so over the following three centuries. Native American men certainly did use mirrors, which were a popular trade item across North America. Because Native American men were more involved in trade with whites, their

Figure 5.3. Gorosimp (Shoshone warrior), between 1884 and 1885. Denver Public Library, Western History Collection, no. 32284.

mirror use would have been more visible to the white men. As a result, white men formed and distributed this association between Native men and mirrors. It served as additional evidence to support the accusation made by whites that Native peoples misunderstood appropriate gender roles. The introduction of mirrors allowed whites to shape an image in white minds of Native peoples as other and inferior.

The question still remains about the extent to which Native American women may have also made use of mirrors. In the earliest encounters between

Figure 5.4. Head-men or chiefs (Arikara), 1872. Denver Public Library, Western History Collection, no. 30933.

Europeans and Native peoples in North America, Europeans sometimes linked Native women with mirror use. It is likely that this assumption arose from the association of women with mirrors in Europe. When an English supporter of New World colonization, Richard Hakluyt, recommended how to outfit an expedition in the sixteenth century, for example, he included "Looking glasses for women, great and fayre." Because Hakluyt never traveled to the New World, he almost certainly acquired his idea of the association of women and looking glasses from his own culture. Roger Williams's assertion

Figure 5.5. Spotted Elk (Sioux), ca. 1908. Western Americana Photographs Collection, 1870–1998 (mostly 1870–1915) (WC064), Box M0825, Western Americana Collection, Department of Rare Books and Special Collections, Princeton University Library.

that it was only women in Narragansett society who used mirrors also likely came from his European background, given that both Narragansett men and women practiced painting and therefore likely made use of mirrors.[16]

Although descriptions of Native peoples using mirrors focused almost exclusively on men, some whites did occasionally note mirrors that were given to or used by women. In 1743, whites in South Carolina provided several material goods for a group of Pedee Indians, including a looking glass for each of the three women. In the long lists of goods that were to be offered on Lewis and Clark's journey (1804–6), most of the looking glasses were designated as gifts for men. Yet there were a few pewter looking glasses earmarked for "young girls." Painter

Figure 5.6. Little Horse, Chief (Oglala Sioux), 1899. Denver Public Library, Western History Collection, no. 31801.

George Catlin (1796–1872) recounted how, during his voyage along the Pacific Coast of North America, he attempted to establish a friendly relationship with a Native man he encountered by giving this man's daughter "a handsome string of blue and white beads" and his wife "a little looking-glass." Among the Nootka on Vancouver Island, Scotsman Gilbert Malcolm Sproat (1834–1913), who lived there from 1860 to 1865, observed that the "women are careful of their hair, and have little boxes in which they keep combs and looking glasses." Together, these fragmentary glimpses of women's mirrors suggest that at least among some Native peoples, mirrors were a desirable item to both men and women.[17]

The only account that provided an explanation in support of the observations repeatedly made by whites emphasizing men's—rather than women's—use

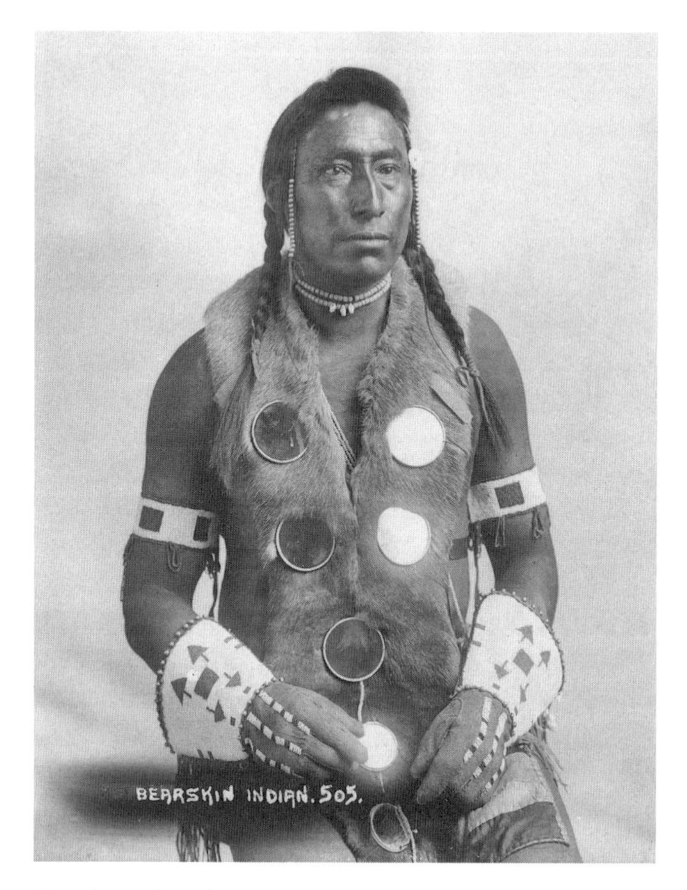

Figure 5.7. Bearskin Indian, between 1903 and 1942. Photograph by Byron Harmon (1876–1942). Whyte Museum of the Canadian Rockies, 505, whyte.org.

of mirrors came from Samuel W. Pond of Connecticut, a white man who moved west and spent "nearly twenty years in intimate association with the Dakota" along the Mississippi River in Minnesota. Pond engaged in missionary work with the Dakota Sioux on behalf of his Christian faith. He repeated the refrain many other white observers made about Native men's mirror use. Dakota men, Pond wrote, "spent much time in painting their faces with various kinds of paints, and carried little mirrors hanging to their girdles, of which they made great use." Native men and women on the whole, however, did not seem to Pond particularly concerned with personal appearance; Pond argued that only "the young men spent much time at the toilet." Perhaps interested in challenging the stereotypical view of Native men with which he must have been familiar,

Pond even defended the masculinity of at least some of the young men. He concluded that even among them, not all "were fops," men who cared too much about how they looked. But then he did something even more unexpected: Pond addressed Dakota women's relationship with mirrors directly. Dakota women, according to Pond, "were told that if they looked in mirrors their eyes would be spoiled." Pond assumed that most of the Native women knew what they looked like nevertheless, observing that "there were other reflectors besides looking-glasses, and probably most of them knew how they looked."[18]

That we know so little about Native women's mirror use is also a part of a larger pattern of Native women's invisibility to whites. As Patricia Albers has argued, European Americans systematically rendered women invisible in their descriptions of Native American cultures. Albers writes that "when White people think of Indian culture, they identify it with the equestrian, buffalo-hunting life of the nineteenth century Plains Indian," and that the "side of Plains Indian life most often seen by the American public is the male half" while Native American women "remain hidden from view."[19] In the case of looking glasses, the invisibility of Native American women was especially meaningful. In reading about encounters with Native Americans and viewing photographs and postcards of Native Americans, whites would have been left with the impression that it was almost exclusively Native American men who used looking glasses. Portraying Native American cultures as ones in which men relied heavily on mirrors, and letting women's absence from the story suggest that they did not, reinforced the idea of a fundamental difference between white and Native American people. Whites imposed their own cultural values and gendered expectations onto Native Americans and found the mirror useful in producing a Native American masculinity that suggested disorder and inferiority. This served their desired ends of constructing white privilege.[20]

While white men claimed use of the mirror to increase their visual mastery of themselves and the world as a tool of rational enlightenment, they assumed, as we have seen, that Native men sought access to mirrors in order to satisfy a feminized vanity. White men did acknowledge, however, that Native men made use of the mirror effectively in one way that did not impugn their masculinity. For their skilled use of mirrors to signal across great distances, nineteenth-century Plains Indians received lengthy praise and admiration from white observers.

Richard Irving Dodge (1827–95), a career army officer who lived in the Plains region for most of his adult life, described witnessing the effectiveness with which a Sioux chief used a looking glass in signaling in his *Our Wild Indians* (1883):

> Once, standing on a little knoll overlooking the valley of the South Platte, I witnessed, almost at my feet, a drill of about a hundred warriors. Their commander, a Sioux chief, sat on his horse on a knoll a little way above me, and some two hundred yards from his command in the plain below. For more than half an hour he directed a drill which, for variety and promptness of action, could not be excelled (I doubt if equalled) by any cavalry in the world. There were no verbal commands, and all I could see was an occasional movement of the right arm. He afterwards told me that he had used a looking-glass.[21]

John MacLean (1851–1928) arrived in Canada in 1873 as a Methodist minister; in 1880 he began nine years of service as a missionary to the Blood Indians across "almost all of what is now Southern Alberta." In his *The Indians of Canada* (1892), MacLean praised Native skill using the looking glass to signal:

> During the troubles in Colorado, between the American soldiers and the Indians, they [the Indians] were thoroughly conversant with the plans of the military, and wherever danger presented itself they were able to keep several miles in advance of their foes. Such apparent activity and keen sightedness was due to the telegraphic communication kept up between the tribes. The small looking-glass invariably carried by the Indian in his native state, is held toward the sun, and the reflection of the sun's rays is directed toward the persons intended to receive the communication. By this means a message can be sent from bluff to bluff, and the sentinels placed there can converse with each other. I have been aroused from my writing desk by the flash from a looking-glass carried by an Indian two miles distant. One day in camp, an Indian's presence was desired, but he was fully two miles away, riding on his horse. A man standing near, took out his glass and with a single movement of his hand the rider suddenly turned on his horse and after a moment's thought rode toward us.[22]

Not everyone was impressed by Native men's use of the mirror in military applications. James Hobbs reported that Comanche men used mirrors "to dazzle the eyes of an enemy taking aim at him, and thus cause his shot to go harmlessly wide of its mark," which Hobbs stated was "rather ludicrous."[23] Yet even Hobbs did not question the efficacy of this technique. The flash of the mirror served an eminently practical purpose as a long-distance communication device or a defensive weapon that helped nineteenth-century Native men master their environment in ways familiar to white observers.

A very different occasion at which a white person might praise a Native person's use of the mirror was when that mirror became a sign, to the observing white, of a Native person's successful assimilation into white society.[24] For whites this meant that mirrors were no longer portable devices associated with men but pieces of household furniture that had moved into their proper sphere of association with women. At the end of his mid-nineteenth-century (1851–57) six-volume government-commissioned report on Native Americans, Henry Schoolcraft reflected on the future of Indian tribes in the United States, specifically on the process of "civilizing" Native Americans. One element of the process of civilizing was material. Schoolcraft wrote that as Native Americans became civilized, "the rude Indian tripod is replaced by well-made chairs and tables; cast-iron stoves, for cooking purposes, are introduced; then a chamber, or a parlor looking-glass, and perhaps a clock." Suddenly, in Schoolcraft's description, the looking glass was associated with a parlor—the quintessential nineteenth-century white feminine space—and with "evidence of female taste in furniture."[25] In a similar description of Native Americans living near Mackinac, Michigan, Schoolcraft described the many signs of civilization among them. In the "best houses," one could find looking glasses, part of that constellation of goods that bespoke a civilized life from Schoolcraft's white perspective: "They are, of late years, temperate, and of industrious habits. . . . They live in substantial houses of squared logs, furnished with good roofs, floors, doors, chimneys, and glass windows. In the best houses are to be seen cast-iron stoves, obtained by purchase at Michilimackinac, together with chairs, small looking-glasses, and bedsteads. . . . Leather shoes have, to some extent, superseded moccasins; and hats are uniformly worn. Years have now elapsed since feathers or paints have appeared as articles of personal decoration."[26] In addition to embracing many white living standards, Schoolcraft noted that face painting was no longer being practiced. These Native peoples had discontinued this practice; thus, the looking glass was no longer associated with it. Rather, the mirror had become one of several markers of Native American acceptance of European conceptions of civility. Moreover, Schoolcraft clearly positioned the mirror as a piece of furniture, lodging it between chairs and bedsteads in the description, although he specifically noted that the looking glasses were "small." As a piece of furniture, the looking glass was no longer a portable object that could be carried with Native men and used, from Schoolcraft's perspective, improperly.

The looking glass was one object—among many other beliefs, practices, and material goods—through which whites established their belief in Native American difference and inferiority. In a centuries-long process that scholars across many disciplines have explored, successive generations of European Americans gathered what they believed to be evidence of Native American inferiority in order to justify the annihilation of their populations and their removal from desirable lands. To white eyes, Native Americans were "savages" whose ways needed to be transformed and whose land would be better used by "civilized" men. Whites anticipated that Native Americans would not understand the proper use of many kinds of material goods and sought out—perhaps even overemphasized—evidence to support this claim. Because most whites did not have extensive (or even any) encounters with Native Americans, it was often sufficient to create—through written accounts, drawings, and later photographs (sometimes reproduced as postcards [see fig. 5.7])—an image in white minds of Native American difference and inferiority. Mirrors played a role in the process through which whites came to understand Native Americans as "other."

Mirrors also reinforced the racial hierarchy that Europeans used to separate themselves from Africans in the New World. Occasionally whites believed they saw evidence of African inferiority in mirror usage. At the very end of the eighteenth century, Médéric-Louis-Elie Moreau de Saint-Méry wrote about the arrival of newly enslaved Africans to Haiti. According to him, while the "natural products of the island" did not surprise the Africans, they were unfamiliar with many of the European goods they encountered in Haiti. Of all of these goods, Moreau de Saint-Méry believed, the "most striking to them [Africans] are the mirrors and the reflections they produce." He described an African who was unable to recognize himself when confronted, for the first time, by his own reflection in a looking glass. While encountering one's mirror self for the first time as an adult was undoubtedly disorienting to all who had that experience, Moreau de Saint-Méry transformed that shared, initial moment of surprise into an assumption about African inferiority by suggesting that Africans never came to understand what they were seeing in the mirror. The African, he wrote, "looks at a mirror, feels the glass, and runs around back to try to find the other copy of himself." When he had convinced himself that there was no one behind the mirror, he then, according to Moreau de Saint-Méry, "performs a thousand antics and makes a thousand faces, trying to imitate the other person." How the mirror worked and what he saw reflected in it was something, Moreau de Saint-Méry concluded, "nobody can explain to him."[27]

The British actress Frances "Fanny" Kemble, visiting her American husband's coastal Georgia plantation in the 1830s, made a more subtle criticism of an enslaved African American woman she encountered there who owned a mirror. Like most white Americans, Kemble linked mirror ownership to the presentation of a public self and was disappointed, therefore, to see that the mirror's owner was "peculiarly untidy and dirty, and so were her children." Kemble recorded in her diary that she "felt rather inclined to scoff at the piece of civilized vanity [the mirror], which I should otherwise have greeted as a promising sign."[28] Although Kemble did not doubt this woman's ability to recognize her mirror self, as Moreau de Saint-Méry had done, she nonetheless questioned her ability to use the looking glass "properly."

These kinds of white criticisms of African Americans' encounters with their mirror selves were, however, quite rare. If whites had seen African Americans using mirrors in ways that they deemed "other," there is little reason to doubt that whites would have mentioned this, as they eagerly emphasized the differences they believed they observed in Native peoples' mirror use. This silence more likely suggests that people of African descent did not frequently use mirrors in ways unfamiliar to whites, at least not in their presence. But the mirror nevertheless played an important role in how at least some people of African descent constructed identity under a slave regime.

The interviews that African Americans gave to employees of the Works Progress Administration, as part of the Federal Writers' Project in the 1930s, in which the interlocutors recounted their childhood memories of slavery, and other nineteenth-century remembrances, are an invaluable resource for understanding how the mirror served the goals of whiteness in antebellum America. The Federal Writers' Project interviewed more than twenty-three hundred formerly enslaved men and women in seventeen states between 1936 and 1938. These firsthand accounts from elderly men and women who were enslaved as children and adolescents in the period just before and during the Civil War provide a unique window into race relations, life under slavery, and memory.[29]

The WPA interviews provide valuable information on everyday life, including material culture. In the original twenty-question script developed by folklorist John Lomax, who worked for the Federal Writers' Project, interviewees were asked specifically about several different types of material culture: the "quarters" in which they lived, the beds on which they slept, the clothes and shoes they wore, and any medicines or charms that had been used to ward off sickness. Because interviewees were not asked specifically about other material possessions, their comments on additional items can be seen as an indication of the importance of those items to them, or at least to the story they were then telling. In this way, as historian Emily West has argued, the interviews "permit the documentation of significant or memorable experiences."[30]

Depictions of encounters with the mirror self allow us a glimpse at how enslaved people of African descent constructed their identities in mid-nineteenth-century America. African American identity was contested terrain under slavery between slave owners and the enslaved. Whites had created a racial hierarchy in America that equated blackness with deprivation and suffering. Enslaved men and women knew that their bodies were often not their own: they could be sold away at any time, their bodily labors produced profit for their masters rather than for themselves, they faced brutal punishment, and women faced the horror of rape by white men and the prospect of bearing children that would perpetuate this cycle. As the anthropologist John Michael Vlach has argued, in the face of the dehumanization to which whites subjected them, "each day the profound human need for some measure of control over one's destiny was deeply felt, resulting to some extent in the slave's living a double life—one for his master's expectations and one for himself."[31] When people of African descent stood before a mirror, the image of the self that met them created another site at which this contest over enslaved bodies and selves took place. The conflict African Americans faced between this bifurcated sense of self—their identification as fully human versus white society's assessment of them as inferior—emerged in encounters with the mirror self.

The battles fought by this bifurcated self wounded African Americans and led some to evaluate themselves by the standards of white society. When South Carolinian Charlie Davis sat down for his WPA interview, he began by addressing directly the racial inequality and tension that was present in many of these conversations when the interviewer was white. His remarks to the interviewer both suggest African Americans' suspicions about why whites would want to interview them and the social conventions that continued to constrain interracial encounters some seventy years after slavery had ended. He told his interviewer, for example, that he had wondered what he "wanted to talk . . . 'bout since I fust heard you wanted to see me," but he then followed up by saying that it was an "honor for a white gentleman to desire to have a conversation wid me." Davis then asked his interviewer if he wasn't the "blackest" man the interviewer had "seen for a season." Although Davis emphasized his own pride in his dark color, he also acknowledged that America's racial hierarchy privileged whiteness over blackness. He concluded, "If I forgits I is dark complected, all I has to do is to look in a glass and in dere I sho' don't see no white man."[32] Davis embodied the struggle of the bifurcated self to resolve his inner pride in his identity with whites' disdain for his ancestry and personhood. In encounters with his mirror self, Davis found a stark reminder of his blackness and its meaning in the United States.

Two other African Americans reported exactly the same sentiment. Georgian Bob Mobley described his childhood under slavery, sleeping "in a little trundle bed right in the room with my marster and mist'ess. Then if they needed anything I'd be right there to get it. An' I didn't want for nothin'." Specifically, Mobley recalled needing a new suit. He told his mistress, and within a couple of days he had the new clothes. But, as Mobley recalled, if these kindnesses ever tempted him to forget his status, his master's house contained a reminder. As he put it, "I'd think I was white 'till I looked in the glass."[33] Similarly, in 1873, a group of students from Hampton, a Virginia school for formerly enslaved African Americans—it was "one of the early success stories in the struggle to educate former slaves in the South"—were on their "fall campaign, giving concerts every evening." One of the students wrote of the friendly reception they received at Mt. Holyoke Female Seminary: "Here we were treated with all the respect and had all the attention paid to us that could be wished or desired. Indeed, one wouldn't think that he was colored unless he happened to pass before a mirror, or look at his hands."[34]

The similarity in these turns of phrase is telling. Mirrors seem to have done more than simply report neutral visual information about the mirror self to African American observers. When African Americans saw—and judged—their mirror selves by white standards, their vision became an instrument of white domination. Scholars have given a name to the process by which members of an oppressed group judge themselves by the standards of the dominant group: "internalized oppression." The oppression that people of African descent in America experienced for centuries "can lead individuals to internalize the messages of inferiority they receive about their group membership. . . . Over time, internalized oppression can become an unconscious, involuntary response to oppression in which members of oppressed groups internalize the negative stereotypes and expectations of their group based on messages they have received from the oppressor."[35] When African Americans who had internalized their oppressed status in the American racial hierarchy stood before the looking glass, they judged themselves through the "white gaze."[36] In much the same way that white women internalized the male gaze, African American men and women internalized a white gaze, which sought out whiteness—as the male gaze sought out beauty—as most desirable. Blackness was considered inferior to whiteness.

Scholars have considered the extent to which an oppressed group can appropriate the material goods of a dominant group and define the meaning of those goods for themselves. In *Down by the Riverside: A South Carolina Slave Community*, historian and folklorist Charles Joyner begins his analysis of the material conditions of slave life by considering the importance of the "material environment" in "not only reflect[ing] the cognitions and perceptions of the slaves" but in also shaping their behavior. Joyner advocates moving beyond a study of "artifacts alone" to a consideration of "the slaves' material culture—that is the knowledge that enabled them to produce and appreciate material things." In North America, Joyner argues, the material culture of enslaved men and women can best be understood through a process he calls the "creolization of culture." African, not white, beliefs directed the meanings associated with and uses for material goods under slavery. Through the creolization of culture, African beliefs became the "grammar" that provided the framework in which material objects—the "language"—were understood.[37] Africans in North America certainly did follow this path in practices that will be described in chapter 6. But here, especially in the case of white-owned mirrors hanging on the walls of white homes and institutions, looking into a mirror and observing one's mirror self does not seem to have been such an act of appropriation. Rather, mirrors could serve as instruments of domination as they reinforced the racial hierarchy whites had constructed in North America.

W. E. B. Du Bois coined the term *double-consciousness* in the early twentieth century to describe the bifurcated self, how an African American "feels his twoness,—an American, a Negro; two souls, two thoughts, two unreconciled strivings; two warring ideals in one dark body, whose dogged strength alone keeps it from being torn asunder." When Du Bois wrote about double consciousness, he described it as the "sense of always looking at one's self through the eyes of others, of measuring one's soul by the tape of a world that looks on in amused contempt and pity." Sometimes, it would seem, the "eyes of others" took the form of the mirror.[38] Cureton Milling, a former slave in South Carolina, remembered the shock he felt the first time he saw his reflection in a looking glass, bought for him by Roxanna, the woman he hoped to marry. He recalled, "I say to her when I takes a look in it, 'Who dis I see in here?' She say 'Dat's you, honey.' I say: 'No, Roxie, it can't be me. Looks like one of them apes or monkeys I see in John Robinson's circus parade last November.' Dere's been a disapp'intment 'bout my looks ever since, and when my wife die, I never marry again." In the looking glass Milling was startled to see the image of something that he had been taught looked like an animal. His mirror self reflected back to him the belief held by many whites that enslaved men and women were somehow less than fully human.[39] At that moment—when Milling saw his mirror self for the first time—the identity he had of himself as fully human came into conflict with white society's view of the meaning of his life under slavery. Moreover, Milling linked this moment of first seeing himself in the mirror to his not marrying again after Roxie's death. This suggested how poorly he came to value himself after becoming acquainted with his mirror self. Milling learned to look at himself with the "amused contempt and pity" that Du Bois would later describe. He saw himself as unworthy of another woman's love, struggling and ultimately failing to reconcile his humanity with his otherness in the American racial hierarchy.

For other African Americans, mirror encounters revealed another bifurcation of self: visible white ancestry. As Henry Clay Bruce detailed it in his *The New Man: Twenty-Nine Years a Slave, Twenty-Nine Years a Free Man*, Africans in America "would have been pure black, were it not that immoral men have, by force, injected their blood into our veins, to such an extent, that we now represent all colors from pure black to pure white, and almost entirely as a result of the licentiousness of white men, and not of marriage or by the cohabitation of Colored men with white women."[40] For slaves who visibly displayed this white ancestry, their mirror selves were a reminder of the atrocities their foremothers had suffered.

In writing about his experiences while enslaved, William O'Neal recalled a pivotal moment from his childhood when he first understood his status as a slave. During his childhood, the Roberts family owned O'Neal and his mother, Laura. O'Neal described his mother as "copper-colored" and himself as a "quadroon." One night at the dinner table, Mrs. Roberts and her husband were praising the work of Laura and William. Of William, who was only eight,

Mr. Roberts said: "That boy William is a perfect prodigy for his age. No professor of mathematics could be more exact than he is. My orders once given never have to be repeated, like a machine he continues his work until it is complete." This conversation took place just before it was time for a new school year to begin, and the Roberts children—Lee and Mary—began to leave "the front gate morning after morning with their satchels well filled with books, and night after night they studied their lessons well, [and] all this could not fail to pass unnoticed by the bright eye of William."

O'Neal wondered why the Roberts children went to school when he did not. One day, soon after the school year began, O'Neal was left alone in the house, and he stood before one of his master's mirrors. Knowing that he was about the same age as Lee and that they were both boys, O'Neal searched his mirror self for some sign of difference but found none. Looking into the mirror, he observed, "Yes, I'm white just like Lee." Eventually, O'Neal asked his mother why he was not allowed to attend school like the other children in the house. His mother explained to him that it was his status as a "slave's child" that determined his fate.[41] What O'Neal saw in the mirror could not be trusted, for his reflected image told him that he was the same as the other children in the Roberts house; the reality of his lived experience told a radically different story.

Like William O'Neal, Alexander Robertson was an enslaved man who had white ancestry. Although he had frequently been told that his father was a white man, no one would ever reveal to him his father's identity. After freedom, Robertson set out to uncover the identity of his father. He asked around and had the opportunity to "'pear in de lookin' glass." Encountering his mirror self, he acknowledged that he knew he was "half white for sure." From these sources of information he concluded that he "was a Robertson, which have never been denied."[42]

O'Neal's and Robertson's mirror images were clear reminders of the unstable relationships underlying race, slavery, and color in the Old South. Based on their mirror images, O'Neal appeared to himself to be white, and Robertson judged himself to be "half white." In O'Neal's case, his white appearance might have been enough to liberate him from slavery had he been in a place where no one knew him. But his mother's established status as a slave enslaved him as well, no matter what he looked like. For visibly white enslaved men and women, having access to their mirror selves was a necessary first step toward passing into white society, if they desired—and were able—to do so. Without access to the mirror self it would have been very difficult for someone like O'Neal to be confident that his African American ancestry was sufficiently hidden to fool white observers. But even intimate knowledge about their mirror selves, as O'Neal's experience makes clear, could not overcome the less visible (or invisible) ancestry that defined him as enslaved in a place where his lineage was known. Although O'Neal could not escape his status as enslaved as long as he remained near people who knew his lineage, in the antebellum era some visibly white men and women with African ancestry did have increasing opportunities to pass into the anonymous environments afforded by American cities.[43] The ability of some people of African descent to pass fully into white society threatened the meaning and claims of whiteness itself. Not surprisingly, as the relationship between physical appearance and race became increasingly contested in the antebellum era, whites began to seek ways to defend their status and sought to prove that the differences between the races were not only real but also observable.[44]

One manifestation of this concern during the antebellum period was an attempt by white women to emphasize the whiteness of their skin, not only as a mark of beauty but also, as historian Mary Cathryn Cain argues, to "dissociate themselves from the degradations of slavery."[45] White women voiced concern when their skin color darkened, and they also sought acceptable ways to artificially enhance the paleness of their complexion so that there could be no question about their whiteness. Both Louise Clappe (1819–1906) and Mollie Sanford (1838–1915)—who moved west with their families at midcentury—were disappointed to find that too much time outdoors had darkened their complexion. In 1851, New Jersey native Louise Clappe arrived with her husband, Dr. Fayette Clappe, at the log cabin that would be their California home. Clappe noted before the looking glass that her time in the sun had made her face "some six shades darker than usual." Newly married Mollie Sanford discovered the same transformation in 1860 after a year spent in Colorado with her husband. She "raised [her] eyes to look in the glass" and found that she did "not look as fair as . . . one year ago. Mountain air has given [me] a browner tinge."[46] White women revealed their awareness of the acceptable parameters of whiteness when they expressed concern that they had begun to fall outside its boundaries.

While white women limited their time in the sun whenever possible, they also found novel ways to enhance their pale complexions. White women eschewed face paints that might create "artificial whiteness" out of the fear that nonwhites might also use them to try to claim a full measure of whiteness. These same women were, however, more than willing to use remedies deemed acceptable to enhance the paleness of their appearance. Such enhancements might be achieved through the application of rouge, which "accentuated the skin's natural whiteness." White women could also use a popular blue pencil "to trace veins on their temples and foreheads." The enhanced blueness of these veins contrasted with the whiteness of their skin to highlight the paleness of the complexion. To create the desired natural look, women who relied on rouge and blue pencils also required access to a mirror to ensure that these enhancements had worked but had not created an artificial appearance, which was discouraged.[47] For white women the mirror served both as an aid to emphasize their white identity with cosmetics and as a judge of whether their skin tone (either natural or enhanced with these techniques) fell within the acceptable parameters of paleness for a white woman.

In the early American context the mirror was a popular object among individual whites, Native Americans, and African Americans, but it also contributed to that "piecemeal wall" of whiteness Bridget Heneghan described as "built from everyday things." African Americans saw themselves through the white gaze. White women attempted to hedge their whiteness with acceptable remedies to enhance it. Whites saw and judged Native American mirror use. All made visible the constructed nature of whiteness, even as it attempted to appear natural. When African Americans used mirrors to pass into white society, they showed that such a piecemeal wall could be breached. But mirrors did more than just fashion whiteness in the lives of early Americans. As the next chapter explores, Native Americans and African Americans incorporated mirrors into ritual life in ways that creolized this item of material culture—making mirrors an object produced by Europeans but imbued with meaning by its users.

CHAPTER SIX

Mirrors in Black and Red

From the thousands of mirrors that poured into Native country from the earliest days of European contact to the small fragments of looking glasses that enslaved people of African descent kept in their dwellings, generations of men and women attested silently to their desire for this object. This chapter attends to that silent witness by attempting to unravel some of the different reasons mirrors were desirable to people of Native and African descent in North America. As we have already seen, access to an accurate reflection of the self shaped how Native Americans and African Americans constructed their identities. But the desire to access reflective glass did not end with identity formation. This desire also arose from uses for mirrors that people of Native and African descent had developed over centuries related to harnessing the power of the sacred with reflective materials.

The practices described in this chapter related to harnessing the energy of the sacred were rooted in West and West Central Africa (later brought to North America) and in the North American Plains.[1] By "practices" here I include the widest possible range of activities that people engaged in to tap into power beyond the physical world. Some of these we might best describe as religious rituals, while others might be better grouped as superstitions; still others were customs that may have been performed even after people stopped believing in their power. Practitioners found European mirrors useful tools in this range of practices and incorporated them appropriately. These practices highlight the significance with which people of Native and African descent imbued reflective surfaces long before Europeans introduced glass mirrors. Among neither group did the mirror radically transform their practices. What mirrors did do, however, was to increase the availability of reflective materials for use in harnessing the power of the sacred. As such, mirrors likely reinforced the role of reflection and helped to sustain, perhaps even to strengthen, the importance of reflection in these practices.

Evidence about the practices described in this chapter come mostly, and most unfortunately, from observations recorded by white observers. Acknowledging the power of reflection in these kinds of practices is something that whites downplayed as part of their own experiences but were keenly interested in recording to document "otherness." Although we caught a faint glimpse of white practices—some of those troubling visions explored in chapter 4—whites by and large emphasized their use of the mirror as a rational tool that extended the limits of human vision and gave whites accurate and useful information about themselves and their world. As early American whites sought out evidence of Native American and African American inferiority to justify the extermination and subjugation of these peoples, they looked for evidence of precisely those kinds of practices they downplayed among themselves. Emphasizing only their own rational use of the mirror, they called attention to what they deemed to be "inferior" mirror practices they observed among people of African and Native descent. Whites attempted to widen the gap they had created between themselves and those they had deemed "other" in North America. The evidence presented in the first part of this chapter must be read through this lens; wherever possible, additional evidence from the practitioners themselves and their material culture has been included in the discussion as well.

Harnessing the Power of the Sacred: West and West Central Africa and North America

The region of West Central Africa, from which 23 percent of enslaved Africans in North America originated, was home to the BaKongo people, "a cluster of ethnic groups who spoke the KiKongo language, who shared a cultural system called the BaKongo and who inhabited the area referred to historically as Kongo. That geographic area . . . consisted of territories now located in the nations of the Democratic Republic of Congo, Gabon, the Republic of Congo, and Angola."[2] Reflective and glittering objects figured prominently in the BaKongo understanding of the world before European trade mirrors arrived in their midst. BaKongo beliefs about and uses for European mirrors in ritual life can be observed on both sides of the Atlantic.[3]

The BaKongo understood the universe to comprise two worlds: one inhabited by the living (with black bodies) and another, located beneath the world of

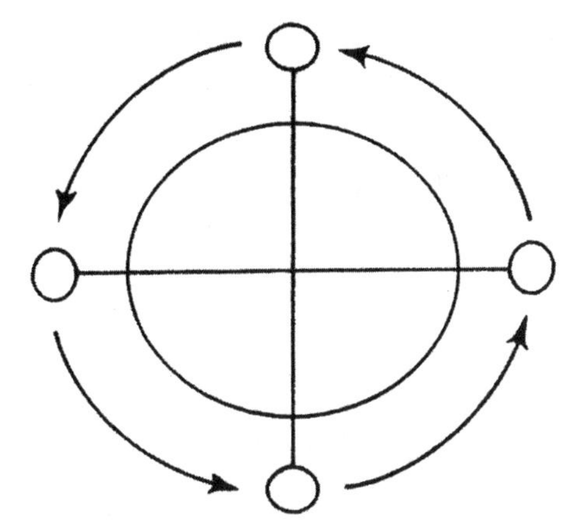

Figure 6.1. Dikenga dia Kongo, the BaKongo cosmogram. From Christopher C. Fennell, "BaKongo Identity and Symbolic Expression in the Americas," in *Archaeology of Atlantic Africa and the African Diaspora*, ed. Akinwumi Ogundiran and Toyin Falola (Bloomington: Indiana University Press, 2007), 203.

the living, inhabited by the dead (with white bodies). These worlds were divided by water, which, in its capacity to act like a mirror, made each world "a reflection of the other." The BaKongo represented their belief about the universe in a cosmogram in which the horizontal line represented *kalunga*, the watery divide between the worlds (fig. 6.1). As the mid-twentieth-century anthropologist Anita Jacobson-Widding has shown, the BaKongo cosmology was more complicated than this clear-cut dualism first appears because "the Congolese midline consists of water," making the boundary between the worlds fluid and creating "the potential for mediation between" the worlds.[4]

The BaKongo believed in a "supreme Godhead," Nzambi, who was "viewed as a remote creator, uninvolved in the daily affairs of the living." They also believed that "Nzambi created a variety of intermediary spirits, known as *basimbi*," who interacted with the living and were associated with the kalunga, that watery divide between the worlds. Basimbi were further associated with reflective and glittery surfaces beyond their connection to the reflective kalunga. G. Cyril Claridge, an early twentieth-century missionary who lived among the BaKongo for twelve years, described the BaKongo basimbi as "fairies" and noted that "when the sun shines on distant objects, such as stones in the hillside, opal, glass, flint, quartz, lakes, etc., their reflection is seen a long way off and is thought to be the clothes of the fairies hanging out to dry after washing."[5]

Just south of the BaKongo lived the Mbundu people, who shared this belief in water spirits, which they called *kianda.*[6] Among the Mbundu we find additional evidence of the association of water spirits with glitter and reflection. The Mbundu believed that kianda dwelled under the water's surface near "great stone formations" that jutted out of the water and were visible from land. These stone formations frequently contained materials that shimmered or glistened in the sun, including quartz and mica.[7]

Among the BaKongo, ritual specialists, known as *banganga*, created *minkisi* (sing. *nkisi*), portable physical containers that "summoned and focused" basimbi to act on behalf of the living. Minkisi could take a variety of forms, among them bowls, gourds, bags, boxes, pots, snail shells, horns, and wooden sculptures. Inside and on the outside of these containers, banganga placed medicine (called *bilongo*), which was key to the object's power. Taken in its entirety—the container and its medicine—an nkisi was "a metaphorical expression of the action to be expected of the spirit." Minkisi also frequently evoked the BaKongo cosmogram either by their creation and use in a "ritual space demarcated by crossed lines drawn upon the ground" or by the image of crossed lines appearing on the body of the minkisi.[8]

Swedish Christian missionary Karl Laman did extensive fieldwork from the 1890s to the 1910s among the BaKongo people. Laman amassed not only a large collection of minkisi but also "indigenous texts describing their origins, composition, uses and ritual context."[9] Minkisi makers incorporated a wide range of medicines into their creations, including objects with reflective qualities. According to Laman, before the introduction of European glass and mirrors, minkisi might contain "beetles or other animals with a metallic lustre." The nkisi in the Laman collection known as Musansi A Bitutu, for example, "treats epilepsy, a disease associated with possession by water-spirits and with the movements of whirlpools." Musansi A Bitutu is a leather bag decorated on the outside and containing objects inside, including "[shiny] pieces of the mbungu ya mputu beetle." The beetle pieces were placed inside this nkisi, perhaps in part a practical choice to protect a fragile item that would not likely survive on the outside.[10]

Laman argued that these fragile lustrous items were later "replaced by bits of mirror or ordinary glass."[11] That Laman claimed nkisi makers used "bits of mirror or ordinary glass" as replacement for beetle wings suggests that these objects (beetle wings, ordinary glass, mirror glass) were desired for their glittering properties—like the shining rocks on the formations jutting out of the water that signaled the presence of water spirits—and not for an ability to reflect the image of something or someone put in front of them (which would have privileged the mirror fragments over the beetle wings and ordinary glass). This glittery property was described by Robert Farris Thompson (b. 1932), an art historian and early expert on minkisi, as "the flash of the spirit." According to Karl Laman, the beetle wings, and the bits of mirror or glass that later replaced them, were intended to "scare away the *bandoki* [witches and wizards] and *bankuyu* [ghosts of former witches, wizards, and sorcerers] with their flash and glitter." Thompson later broadened the interpretation by suggesting that the glitter might "capture, attract, or repel a spirit, and to suggest the flashing waters through which one peers into the other world."[12]

Minkisi also took other forms. Wooden-statue minkisi used small, whole mirrors or pieces of mirrors on the outside, covering the medicine, which was also attached to the outside of the statue. One such nkisi in the Laman collection is *Mbongo Nsimba*. This wooden male figure has a round "medicine pack" attached to its belly (there is also one missing from its head); facing out on the top of the belly medicine pack is a mirror large enough to see a reflection.[13] Another nkisi, this one in the collection of the Brooklyn Museum, also contains a mirror in the statue's abdomen (fig. 6.2). One specialized type of wooden-figure nkisi was the *nkondi* (plural *minkondi*), or nail fetish. Minkondi have nails driven into them, an act that is "intended to anger the nkisi but also represents the suffering that it will then inflict upon its designated victim."[14] Nkondi might also have abdominal medicine packs covered by mirrors (fig. 6.3). Other kinds of minkisi containers might have mirrors attached to the outside as well. The Nkisi Matenzi, for example, used for healing, is a snail shell with a mirror affixed to it.[15]

These larger, visible reflective surfaces (compared to bits of mirror glass or other shiny materials that might be placed inside or on the outside of an nkisi) enabled one to see a reflection in them and thus may have been used for different purposes. Héli Chatelain (1859–1908), a protestant missionary and linguist in Angola in the late nineteenth century, claimed that "all the fetish-images of the Kongo nation wear, incrustated on the stomach, a piece of looking-glass" and were used in divination to show "the face of the culprit." Chatelain saw the similarity between what he observed and European practices as well. He noted that the "speaking mirror, or a mirror revealing secrets, occurs in Portuguese and other tales, and is to this day to be seen for money in European country fairs, where many educated lovers consult it with as much credulity as the

Figure 6.2. Kongo. *Male Figure with Strips of Hide (Nkisi)*, nineteenth century; wood, hide, glass mirror; 14 1/2 × 12 1/2 × 13 in. Brooklyn Museum, Museum Expedition 1922, Robert B. Woodward Memorial Fund, no. 22.1455.

African consults his doctors."[16] As we observed in chapter 4, there is evidence that people of European descent in North America also continued to make use of mirrors in these practices late into the nineteenth century.

Larger mirrors may also have been used to evoke the kalunga. Robert Hamill Nassau (1835–1921), a Presbyterian missionary in West Central Africa for forty years, observed the construction of what he called a "fetich amulet." This particular nkisi was made using an animal horn that held "the various articles deemed necessary to attract and please the spiritual being whose aid is to be invoked." The process of mixing the items that would go into the horn was done in secret. It was accompanied by "public drumming, dancing, songs to the spirit, looking into limpid water or a mirror, and sometimes with the addition of jugglers' tricks, e.g., the eating of fire."[17] That

Figure 6.3. Kongo (Solongo or Woyo subgroup). *Power Figure (Nkisi Nkondi)*, late nineteenth to early twentieth century; wood, iron, glass, fiber, pigment, bone; 24×6 1/2×8 1/2 in. (61.5×17.0×21.5 cm). Brooklyn Museum. Gift of Arturo and Paul Peralta-Ramos, 56.6.98.

limpid water or a mirror could be used here suggests an interchangeability between the two. These reflective surfaces may have been intended to evoke the kalunga because an nkisi would summon one of the basimbi who dwelt there.

Evidence from the late twentieth century is also suggestive of how mirrors came to represent the kalunga. The Swedish anthropologist Anita Jacobson-Widding did fieldwork in the Lower Congo, West Central Africa, intermittently between 1966 and 1974. Near the end of this period, Jacobson-Widding learned of a woman whose husband had died in a car accident. The widow "was secluded together with her sister" for most of a year. On the day that she would reenter society, the woman received new clothing, makeup (lipstick, light beige foundation, and rouge), and a looking glass, which the widow's sister declared to be "the most important thing." Jacobson-Widding asked her to explain, which she did: If the widow "'does not have a looking-glass she will not be able to see her face properly,' the young woman answered. 'She will be so beautiful, after bathing in the river. In the old days people did not have looking-glasses. So they had to try to see their face in the river. But it is not smooth enough.'" When the young widow donned the makeup and looked in the mirror, she saw a whitened face looking back at her—from the other world (on the other side of the kalunga), where bodies were white—that "might have been a water spirit, for instance that of her husband. . . . Yet, at the same time, it was her own face." Jacobson-Widding concluded that "the 'Self' may thus be constructed through interaction with the 'Other' in this world and in the world beyond the water as well."[18] Looking in the mirror not only enabled this young widow to connect with the world beneath the water; the mirror itself evoked that watery barrier between the worlds.

Reflection served a variety of important roles outside the BaKongo culture in regions of West Africa and West Central Africa as well. What is today the nation of Ghana was part of the Gold Coast during the era of the African slave trade.[19] Here we find examples of the importance of reflection and how European mirrors came to be seen as another means to access reflective power. As early as the 1660s, Wilhelm Johann Müller described a "comfu" in the Fetu Kingdom—what he understood to be a sorcerer who claimed to be able to predict the future—who could "prophesy by looking continually into a basin full of water and pretending to perceive something most wonderful in it."[20] In the collection of the Pitt Rivers Museum at Oxford University is a small glass mirror accessioned in 1925 that is recorded to have been used in divination in a village sixteen miles from the coastal town of Sekondi, Ghana.[21] Farther inland from the Atlantic coast, the Ashanti people celebrated an annual eight-day period known as Apo, described by William Bosman in 1705 as including "all manner of Singing, Skipping, Dancing, Mirth, and Jollity; in which time a perfect lampooning [of] Liberty is allowed and Scandal so highly exalted, that they may freely sing of all the Faults, Villanies and Frauds of their Superiours, as well as Inferiours without Punishment, or so much as the least interruption." Cultural anthropologist Victor Turner interpreted Apo as one of the "rituals of status reversal." It was also witnessed by early twentieth-century British anthropologist Robert Rattray in 1923. In his description Rattray mentioned Ashanti priestesses who "practise the art of

Figure 6.4. MS 445/3/71. Priest dancing and gazing into a mirror. Photographed by Robert Sutherland Rattray, c. 1923. © RAI. Courtesy of the Royal Anthropological Institute, London.

water gazing," and he included a photograph of a male priest holding up a looking glass (fig. 6.4).[22]

There is also evidence of a belief in the power of reflection among people living along the River Niger in the region of present-day Nigeria.[23] In his 1830s journey along the River Niger from Bussa to the Atlantic Ocean, Richard Lander noted an encounter with a man he described as a "Mohammaden" priest who asked for a specific size and shape of a mirror to be brought back with them from England for his use.[24] At Onitsha, a town located along the River Niger, a Christian minister, J. C. Taylor, a native of Sierra Leone, founded an Igbo mission in 1858. In his journal from Onitsha, Taylor recorded how an Itshi man living "one or two days' journey" from Onitsha had visited them and been given a mirror as a gift. Less than a month later, the same man returned seeking another mirror; he explained that the mirror they had originally given him was broken, and he needed a new one. He told them that "he was a great doctor . . . [who] had long bewitched his countrymen. One day he wanted to exhibit wonders to them," so he showed them this "wonderful reflector." He got it out to show them "and unfortunately smashed it to pieces." Knowing that he was using the mirror to "bewitch" caused the Christians at Onitsha to refuse his request. They told him that "God has given us knowledge to make glass, but does not sanction us to dedicate it to superstitious rites, and to deceive our fellow-creatures; consequently you have lost the glass altogether, and we are sorry we can do no more for you."[25]

Farther southeast, in the interior regions of West Central Africa, along the border between the present-day nations of Congo and Democratic Republic of Congo, the Englishman E. J. Glave founded the Lukolela Station on the River Congo around 1883.[26] At Lukolela Glave reported that the residents there ascribed to him "supernatural power; a belief which I did not correct." They believed his literacy enabled him to see into the future through the books he read. They also, Glave reported, asked him to look into his mirror and tell them "whether a sick child would recover."[27]

West and West Central Africans also widely decorated graves with material goods, including mirrors. Anthropologist John Michael Vlach summarizes the most common elements of this tradition: "the pattern and practice of burial customs across West and Central Africa is roughly equivalent. Graves are marked by common material possessions; they are often broken and must not be disturbed because they are the property of the deceased." Both Glave and Nassau noted the practice of covering the grave with material goods in the late nineteenth century, and Nassau specifically noted that one of those goods was a mirror: "pieces of crockery, knives, sometimes a table, mirrors, and other goods obtained in foreign trade." On grave sites, as Vlach has noted, mirrors can be categorized with a group of objects that may all be water symbols, including seashells and ceramic pitchers that hold water when in use.[28] This grouping of mirrors with other water symbols suggests again that evocation of the kalunga to mark the divide between the land of the living and the land of the dead to which the recently deceased must journey.

The African beliefs and practice of nkisi-making, ritual practices including divination, and grave-site decoration all appeared among communities of African descent in North America as well. In his 1983 *Flash of the Spirit: African and Afro-American Art and Philosophy*, Robert Farris Thompson argued that there was "a surprisingly strong *minsiki*-making [*sic*] tradition" in North America. Written before several archaeological discoveries provided material evidence of minkisi, Thompson found clues in nineteenth-century folktales, in twentieth-century practice, and in his consultations with Fu-Kiau Bunseki, "a leading native authority on Kongo culture." Later archaeological finds confirmed Thompson's argument, including African-made colonoware marked with the BaKongo cosmogram found on riverbeds in the low country of South Carolina and caches of objects discovered at various archaeological sites throughout the South, including the Carroll, Brice, and Slayton houses in Annapolis, Maryland; the Locust Grove Plantation in Louisville, Kentucky; Utopia and Kingsmill Plantations on the James River just south of Williamsburg, Virginia; Eden House, Edenton, North Carolina; and at the Levi Jordan Plantation in Brazoria, Texas.[29]

Mirrors were not included in the caches uncovered at all of these sites, but those at the Utopia Slave Quarter (ca. 1750–80) and the Levi Jordan Plantation site (settled in 1848) had mirrors. At the Utopia Slave Quarter a mirror fragment was recovered along with "a piece of fossil coral emblematic of water," but the artifacts' location at the time of discovery went unrecorded, so it is impossible to know for certain whether the items here represented an intentional cache. At the Levi Jordan Plantation in Brazoria, Texas, archaeologists have studied what they believe to have been a "curer's cabin" based on four groupings of material culture found buried in the floor. "One deposit consisted of a concentration of small iron wedges," which they think may be the only remaining evidence of an nkisi Nkondi. The medicine for this nkisi may have been the "water-worn pebbles, fragments of mirrors, several seashells, and a part of a small white porcelain doll" discovered nearby.[30]

The use of mirrors in a wide range of ritual practices, including divination, appeared in North America as well. Former slave Charity Jones recalled being told, as a young girl, that she could take a piece of looking glass on the first day of May and look over her left shoulder with the sun shining into the glass. If she did this, she would see in the glass the image of the man she was destined to marry. When she held her piece of looking glass into the sun, the man whose image was reflected in it was Sam Jones, who later—as promised—became her husband. Early twentieth-century anthropologist Newbell Niles Puckett's *Folk Beliefs of the Southern Negro* (1926) recorded a similar tradition but added that you might also see "anything else that is going to happen before the year is out, or simply your coffin."[31]

African American conjurers also incorporated mirrors into their practices. Conjurers were important figures among enslaved Africans.[32] Former slave Marrinda Jane Singleton recalled that "conjuration was carried on by quite a few," most of whom were from "the Indies and other Islands" although some of their knowledge had been "handed down from the wilds of Africa." Singleton recalled that although masters disapproved of conjuring, many enslaved men and women continued to believe because they "feared de charm of witch craft more than de whippin' dat de Marster gave."[33] One conjuring practice that involved a mirror was the "black cat bone," which could be "obtained by boiling a black 'boar' (tom) cat." To choose which of the cat's bones is the powerful one, the bones must be taken to a crossroads "in the woods where no one will see you." In addition to the bones, a mirror must be brought. Then one was instructed to "stand directly between the forks with your back to the straight road holding the mirror up before you so that the road behind is reflected. Then hold your mouth open and pass the bones, one by one, through it, looking into the mirror all the time. When you get to the right bone the mirror will become dark—you cannot see a thing in it." Once correctly identified, possession of the black cat bone could produce a variety of outcomes: the ability to become invisible, protection against evil, wealth. When looking glasses were not available (and perhaps before they were available at all), water may have been used to identify the black cat bone. In some cases the bones were put in running water, and the black cat bone was identified by its refusal to sink or by its ability to flow upstream, apart from all the others.[34]

The final category of West and West Central African practices—grave-site decoration—also appeared in North America. As Thompson writes in *Flash of the Spirit*, African-influenced grave decorations "cryptically honor the spirit in the earth, guide it to the other world, and prevent it from wandering or returning to haunt survivors. In other words, the surface 'decorations' frequently function as 'medicines' of admonishment and love, and they mark a persistent cultural link between Kongo and the black New World." Two twentieth-century graves in coastal Sunbury, Georgia, are marked with mirrors. One has a 2 ft. × 3 ft. mirror embedded into the "concrete slab that covered the top of the burial mound." Given the size of the mirror and its horizontal position, John Michael Vlach concluded that this mirror "was apparently a dramatic representation of the watery transition between life and death."[35] The other tombstone marks the grave of Rachel Bowens-Pap, who died in March 1937 at fifty-four years of age. The marker is a stark clay rectangle painted yellow. The only decoration on the gravestone is a handprint into which has been placed a small piece of a mirror (fig. 6.5).[36] Thompson argued that some mirrors incorporated into twentieth-century graves in Georgia were taken from the deceased's home and placed on the grave to keep "the spirit at safe distance from the living."[37] The mirror in the tombstone of Rachel Bowens-Pap may have served this purpose.

Reflective and glittery objects and surfaces—and later European mirrors—played a role in the rich set of beliefs held by West and West Central Africans about the powers of the world beyond their earthly existence and the ways in which human beings could harness those powers in this world. These beliefs manifested themselves in nkisi-making, divination, other ritual uses, and grave-site decoration, all of which were practiced in Africa and North America.[38]

Figure 6.5. Grave of Rachel Bowens-Pap. Photograph by Orrin Sage Wightman, from *Early Days of Coastal Georgia* (St. Simons Island, GA: Fort Frederica Association, 1955), 220.

Evidence from the early twentieth-century WPA interviews speaks to the ways in which mirrors were incorporated into other beliefs and practices. In part, these other beliefs and practices may have developed because mirrors became a more common possession among people of African descent in North America. As mirrors took up residence in many dwellings, the object's power may have dictated that a system of beliefs develop to mitigate its potential dangers.

Many WPA interviewees reported on their beliefs in mirrors to affect human lives. Breaking a looking glass was believed to bring bad luck. William Emmons (Ohio/Kentucky), Cheney Cross (Alabama), Annie Boyd (Kentucky), and Mary Wright (Kentucky) all spoke of the bad luck that resulted from a broken looking glass. Boyd believed that a broken looking glass would bring seven years of bad luck. Emmons and Cross had both been taught that this bad luck could be warded off if the pieces of glass were put into running water. Emmons said that if the mirror was thrown in a "runnin' stream de bad luck would flow away."[39] Yet Mary Wright believed that if one broke a mirror, there was nothing that person could do to "keep from having bad luck. Nuthin you do will keep you from hit." Even more ominous was the belief of Lidia Jones (Arkansas) that breaking a looking glass meant that someone would soon die. "I know I broke a small one and I lost my sister. And another thing—I had a lookin'-glass in my trunk and quilts on top of it at another woman's house, and it got broke and she lost her mother-in-law."[40]

Beliefs about mirrors surrounded the beginning and end of life. There was an overall sense that babies should not be allowed to see themselves in mirrors. May Satterfield (Virginia) included in her discussion of superstitions under "Bad Luck" that "to let the infant see itself in the mirror will make its teeth hard." Newbell Niles Puckett offered a more detailed explanation: "A baby should never be allowed to look into a mirror (before he is a year old) or he will have trouble in teething; will be cross-eyed (especially if he sees his father for the first time in a mirror); or ugly; or, if you stand him before the mirror before he is old enough to talk, he will 'talk tongue-tied' or not at all."[41]

Mirrors were also associated with rituals surrounding the time of death. The belief that the reflective surface of mirrors should be obscured after a death—either by turning the mirror to face the wall or covering it with a cloth (usually white was recommended)—appeared in several WPA narratives. Reasons for these activities varied. A woman identified only as "Mrs. Rush" (Georgia) explained that mirrors had to be covered "or else they will lose their shine and will never be any more good. She then pointed to a picture on the wall and said that the dull appearance of the glass was due to her failure to turn the picture to the wall at the death of her husband." Robert Heard (Georgia) believed mirrors had to be covered to prevent another death in the family. Similarly, Aunt Clussey (Alabama) told how the mirrors had to be wrapped because if the dead person's spirit saw itself in the mirror, there would be another death in the house. It was important to keep the dead from reflecting their images in the looking glass as well because, according to Casie Jones Brown (Arkansas), "if you see a dead man in the mirror, you will be unlucky the rest of your life." Hamp Kennedy (Mississippi) provided an origin story for why mirrors had to be covered at the time of death. Kennedy had heard a story that once "a long time ago" a man had died, and three days after his death, when his family looked into a mirror, they saw the dead man reflected back to them as "plain as day in de mirror." After that time, according to Kennedy, people began to turn the mirrors to the wall to prevent this from happening again.[42] Newbell Niles Puckett was even given material evidence of the damage a corpse had done to a mirror. Puckett wrote that he was told "a reflection of the corpse might permanently hold in either the pictures or mirrors. I have in my possession an old mirror with two defects in the silvering which an imaginative person may conceive of as resembling human eyes. The original owner says, 'Us didn't kivver hit up when May (his first wife) died, an' in jes' a day or so afterwards her eyes popped out on hit.'"[43]

On one occasion the WPA interviewers witnessed someone using mirrors because of a belief in their power. White workers for the Savannah Unit of the Georgia Writers' Project met an African American man, George Boddison, and were immediately struck by his appearance. Boddison's "wrists and arms were encircled by copper wire strung with good luck charms; his fingers were covered with several large plain rings." More copper wire "was bound around his head and attached to this wire were two broken bits of mirror which, lying flat against his temples with the reflecting side out, flashed and glittered when he moved his head." Boddison explained: "'But deze dat I weahs,' indicating the copper wire, the mirrors, and the other charms, 'keeps all deze tings from huttn me. Duh ebil caahn dwell on me. It hab tuh pass on.'"[44] That flash of the mirrors and their deflective capacity, Boddison asserted, kept evil from being able to dwell on him; rather, the evil had to move on past him as the mirrors deflected it away.

These beliefs in powers associated with mirrors were not, however, universally held by African Americans. In the Georgia interviews formerly enslaved men and women were specifically asked these questions: "Were you superstitious? Did you believe that a screeching of an owl was the sign of death? That the bellowing of a cow after dark is a sign of death? That sneezing while eating, or if a dog howls after dark it is a sign of death. Did you believe that if there is a death in the house, the ticking of the clock must be discontinued at once and a white cloth hung over a mirror?"[45]

These specific questions revealed three Georgians who denounced such practices and beliefs in their own lives: Mary Jane Simmons, Elsie Moreland, and Robert Henry. Mary Jane Simmons firmly answered that her mother never allowed her to be superstitious. Elsie Moreland believed some of the things she was asked about, but she remembered covering mirrors only to follow "the example of the old people who had gone before her." Moreland undoubtedly spoke for others when she framed these acts as customs performed in honor of those who had gone before her and who, presumably, had believed in them. Robert Henry claimed he did not believe in any such "foolishment," but he did cover mirrors with a white cloth at the time of death (and stopped clocks) just "to be on the safe side and prevent another death in the family." The Works Progress Administration interviewer recorded that "Uncle Robert's highly educated daughter smiled indulgently on him while he was giving voice to these opinions, and we left him threatening her with dire punishment if she should ever fail to carry out his instructions in matters of this nature." Carrying on these traditions may have, over time, represented less active belief than it did respect for elders and ancestors.[46]

When questioned directly in the 1930s about whether they were superstitious, some elderly African Americans—and their children—from Georgia sought to distance themselves from the beliefs of previous generations. Whites had, as we have seen, frequently denied having such superstitions throughout the nineteenth century, although there is evidence that such beliefs persisted among them. This desire to decrease belief in and reliance on superstition—or at least to deny such beliefs—may have existed among African Americans during the nineteenth century as well. When finally asked for their perspective on nineteenth-century superstitions, one WPA interviewee described both that white denial and the level of belief he believed it masked: "Whites called the slaves superstitious. Yet more whites were superstitious than slaves. The slaves used superstition to fool the white man."[47]

The WPA interviews reveal a wide range of beliefs and opinions about the potential power of mirrors both under slavery and at the time the interviews were conducted. They also document how practices developed to accommodate mirrors as they became a common piece of household furniture. The need to protect a baby from seeing its reflection in a mirror, for example, would have become a necessity only when mirrors were so common and prominently placed that babies would have been frequently in rooms with mirrors and thus in need of that protection. Moreover, the WPA interviews remind us of the complex web of practice and belief that surrounded peoples' interactions with reflective surfaces and how, even in one individual's lifetime, an active belief could be transformed into a custom that kept the outward form of interaction with the mirror intact even as the meaning behind it disappeared, until, finally, even the practice was but a memory.

Harnessing the Power of the Sacred: North American Plains

One of the central features of nineteenth-century Native American Plains life was ritual ceremonies and dances.[48] Plains Indians incorporated mirrors into these practices. The earliest material evidence of this incorporation comes from the 1830s. Dr. Nathan Sturges Jarvis, a U.S. Army surgeon commissioned at Fort Snelling, Minnesota, collected Native American material culture in the 1830s. His collection, now housed by the Brooklyn Museum, New York, includes a Sioux mirror board. Shaped like an extended diamond, this thirty-two inch piece of wood has two cutouts at the wider end: one at the top and the other nearer to the middle of the board. On the wooden surface between the cutouts are two rectangular mirrors embedded into the wood (fig. 6.6). In 1839, Philadelphian Caleb W. Pusey acquired a Winnebago War Club while at Fort Winnebago, Wisconsin. The Winnebago were a Siouan-speaking people located in Wisconsin. The heavily engraved 15 1/2 in. long Winnebago War Club was made from antler (fig. 6.7).[49] These mirror boards show a resemblance to the kind depicted in a painting by the Swiss artist Peter Rindisbacher (1806–34), "the earliest artist of note in the Minnesota country." His family came to North America when he was fifteen years old, and Rindisbacher lived there until his death at twenty-eight, settling first at Pembina, Canada, on the Red River, then at Fort Snelling, Minnesota; then in Wisconsin; and finally St. Louis, Missouri. His ca. 1838 painting *War Dance of the Sauks and Foxes* depicts seventeen men, four of whom, all in the right half of the picture, carry a carved piece of wood, any of which might have been a mirror board similar to the ones collected by Jarvis and Pusey (fig. 6.8).[50]

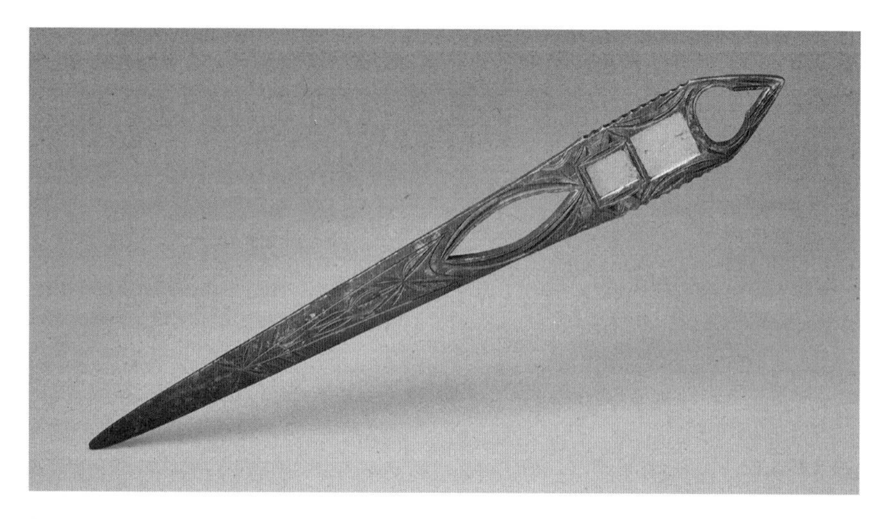

Figure 6.6. Sioux (Native American), carved dance mirror, early nineteenth century; wood, pigment, glass, 32 × 3 3/8 in. Brooklyn Museum. Henry L. Batterman Fund and the Frank Sherman Benson Fund, no. 50.67.96.

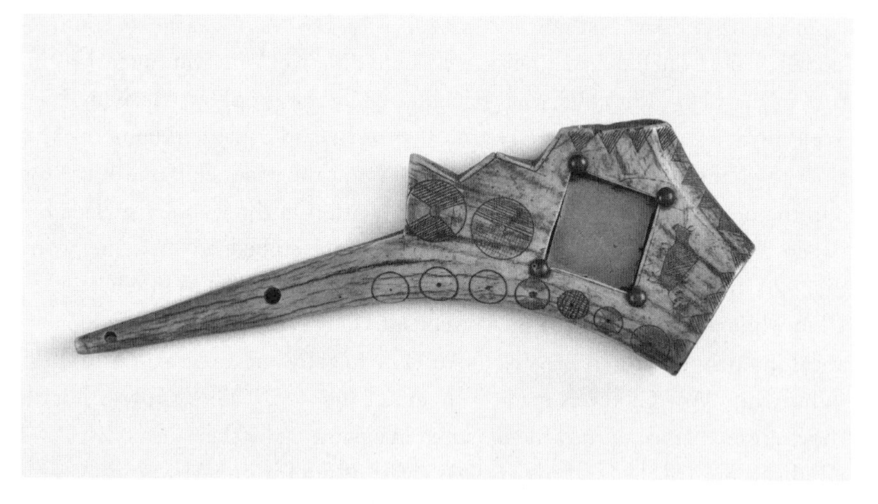

Figure 6.7. Club (Winnebago). Courtesy of Penn Museum, image no. 237306.

While the material culture examples from the 1830s provide evidence that mirrors were incorporated into ritual and ceremonial material culture of the Plains Indians, we can gain further insight into how these items were actually used and the meanings ascribed to them through the work of late nineteenth-

Figure 6.8. War Dance of the Sauks and Foxes. Lithograph by Peter Rindisbacher, ca. 1838. Courtesy of Library of Congress Prints and Photographs Division.

century ethnologists and other whites who studied Plains cultures. Their work provides a context for how objects similar to these, and other mirrored items, were used by these Native American cultures. While some of the ceremonial and ritual uses of the mirror among Plains Indians relied on the object's ability to reflect an image of what was placed in front of it, even more common seems to have been the use of mirrors for their ability to deflect light in a shiny flash or glimmer.[51]

During her 1882 stay at the Pine Ridge Reservation in the Dakota Territory, ethnologist Alice C. Fletcher (1838–1923) witnessed the ceremony of the Oglala Sioux (part of the Western Teton Dakota) Elk Dreamers Society. The elk was an important visionary animal to the Oglala, who saw in the male of the species a "mysterious power" to attract females. Because of this ability to attract, the bull elk became for Oglala men the "incarnation of the power over females."[52] In visions the elk appeared without a heart to signal his immortal and supernatural status. While the elk appeared to individual men in these visions, the Elk Dreamers performed their ceremony after such a vision because the Oglala Sioux believed that "vision powers were dormant until they were publicly performed."[53]

The Elk Dreamers ceremony that Fletcher witnessed began in response to a man in his early twenties who had seen the visionary elk. The initiate invited the members of the Elk Dreamers Society to gather. Once assembled, they prepared the tent central to the ceremony: "Over the place in the tent usually occupied by the fire," they laid down plant material and then placed upon that plant material a "square looking-glass on which lines made of fine dark earth extended from corner to corner, making a cross." The men wore masks "resembling the heads of elk." One of the masks had looking glasses "in place of eyes," and another had "a small circular looking glass like a single eye fastened on the forehead," but almost all of the masks, Fletcher noted, "had something fastened on them which would catch and reflect the light."[54]

After the tent and the participants were prepared, a lengthy ceremonial walk was begun, led by four young women, two of them carrying specially prepared pipes. The men followed these women as they walked, with the men imitating elks with each step taken: the men emerged from the tent, for example, "indicating caution, as the elk might step forth from cover and look about him." A few men carried hoops "containing a square from which depended a fringe of rattling deer hoofs." In the center of only the initiate's hoop was "a circular mirror, fastened by four cords, from which he cast a reflection of the sun from time to time upon the ground, or held up the hoop and flashed the mirror."[55]

These mirror hoops can also be seen in a drawing of the Elk Dreamers ceremony made by an Oglala Sioux from the Pine Ridge Reservation. Between 1890 and 1913, Amos Bad Heart Bull created hundreds of drawings of the Oglala Sioux. In the 1920s and 1930s Helen Blish, a graduate student at the University of Nebraska, aided by He Dog and Short Bull—two elderly Oglala Sioux from Pine Ridge—brought the large collection of drawings to light and wrote accompanying text that described the drawings in detail (Blish, He Dog and Short Bull will hereafter be referred to as the Pine Ridge group). In Drawing No. 115, which depicts the "cultic ceremony of the elk dreamers," three men dance with hoops that have a small, circular mirror at the center (fig. 6.9).[56]

There were, then, according to Alice Fletcher and the Pine Ridge group, at least three kinds of mirrors used by the Oglala Sioux Elk Dreamers Society in this ceremony: the mirror that lay in the tent where the fire would have been, mirrors (and other reflective objects) fixed to the dancers' masks, and the hoop mirror (according to Alice Fletcher, carried only by the initiate; according to the Pine Ridge group, carried more widely in the ceremony). While Fletcher remained silent on the meaning of the mirrors, the Pine Ridge group noted

Figure 6.9. Performance of the Cultic Ceremony of the Elk Dreamers. Reproduced from Amos Bad Heart Bull and Helen H. Blish, *A Pictographic History of the Oglala Sioux* (Lincoln: University of Nebraska Press, 1967), 200.

that "the mirrors, so it is said, possess the power to detect and track ill-feeling . . . where the evil spirit is present to harass one and test his strength and power." The "ill-feeling" in this ceremony, according to the Pine Ridge group, came from the "squatting figure . . . who tests the medicine power of the dancers" (in the lower left of fig. 6.9).[57]

Another interpretation of the meaning of these mirrors was added in the early twentieth century by Clark Wissler, an anthropologist who did not witness an Elk Ceremony but who did spend time with Native peoples in the Plains and wrote about the Elk Dreamers using Fletcher's observations. Wissler's perspective is now widely criticized as part of "salvage ethnography," which sought to record Native American cultures believed to be on the brink of extinction rather than "primarily to understand" those cultures.[58] Wissler argued that when used in the Elk Dreamers ceremony, the mirror "was flashed so that the beam would fall upon the girl . . . [who] must follow the footsteps of the owner of the mirror like the females of his kind follow the male elk."[59] This interpretation is supported by a Dakota (Eastern Sioux) wooden fan circa 1840 that depicts on one side a male elk surrounded by twenty-three female elk. On the other side there are thirty-one female figures around the outside of the fan, and at the center of the fan appears a circular design with a line coming out from the center circle to each of the women's heads—as if the center object were a mirror reflecting a beam of light to each woman. Although they differed on the mirror's meanings, those who wrote about the Elk Dreamers' use of the mirror and the hoop agreed that they were "powerful medicine" and a "very important religious symbol" to the Dakota.[60]

The Oglala Sioux were not the only Plains Indian culture to incorporate mirrors into their ceremonial practices. The Dakota Sioux used mirrors in their Grass Dance. The Grass Dance included a motion that evoked planting corn in the ground; on some occasions the item used to represent the hoe was a mirror, framed in such a way (likely a mirror board, see figs. 6.6 and 6.7) that it could be used to "make" the holes in the ground into which the corn would be placed.[61] And mirrors appeared at least in the visions of the Teton Sioux's Sun Dance, their "most important public religious ceremony." In one account of this dance told to Frances Densmore—an early twentieth-century ethnomusicologist whose work among the Teton Sioux continues to be well-respected—one of the dancers, Red Bird, "had a vision" in which he saw the leader of the dance holding "a small mirror in his hand" from which "he threw the light reflected from this mirror into the face of one dancer after another, each man falling to the ground when it flashed into his eyes. At last Red Bird felt the flash of light in his own face and fell unconscious." Because of this vision, Red Bird was initiated into a leadership role in the Sun Dance.[62]

The Elk Dreamers ceremony, the Grass Dance, and the Sun Dance were complex and elaborate ritual performances. Mirrors were just one object among many used to enact these rites. Moreover, Alice Fletcher's observation that all of the Elk Dreamers' masks contained some type of reflective item (not all of them were mirrors) suggests that, like the BaKongo in West Central Africa, European mirrors may have been incorporated into practices that already involved objects that provided a flash or glimmer.

There is also material evidence for European-made mirrors being removed from their original frames and encased in frames or boards that depicted animals that held key meaning in Native societies. Edwin James was a member of Major Stephen H. Long's 1820 expedition "commissioned to ascend the Platte River and explore the headwaters of both the Red River and Arkansas River." In his account of the journey James described seeing a man carve the image of an important animal for Kiowa Apaches in a new frame. Once this man received the looking glass, a gift from Long's party, he "immediately stripped off the frame and covering and inserted it with some ingenuity into a large billet of wood, on which he began to carve the figure of an alligator." According to James, the Kiowa Apaches held the alligator in high esteem, wearing images of the animal "either as ornaments or as amulets for the cure or prevention of disease and misfortune."[63] Other Plains Indians constructed mirror boards in the shape of horses, another animal of key significance.[64]

Medicine men on the Plains also harnessed the power of reflection to heal. George Catlin's 1835 painting *Blue Medicine* featured Tóh-to-wah-kón-da-pee, a medicine man "of the Prairie Village band of Mdewakanton Santee," painted while Catlin was at Fort Snelling, Minnesota. In addition to a drum and rattle of antelope hoofs, Tóh-to-wah-kón-da-pee wore a prominent circular looking glass framed in metal and hung on a red cord around his neck (fig. 6.10).[65] Medicine men also included mirrors in their medicine bags. Brave Buffalo,

Figure 6.10. Tóh-to-wah-kón-da-pee, Blue Medicine, a Medicine Man of the Ting-ta-to-ah Band, 1835. Eastern Sioux/Dakota. Painting by George Catlin (1796–1872). Oil, 29×24 in. Smithsonian American Art Museum. Gift of Mrs. Joseph Harrison Jr., no. 1985.66.73.

described by Frances Densmore as "one of the most powerful medicine-men on the Standing Rock Reservation" (in North and South Dakota), had in his medicine bag "a small mirror inclosed in a flat frame of unpainted wood, the whole being about 4 by 6 inches. On the mirror was a drawing of a new moon and a star." Brave Buffalo told Densmore, "I hold this mirror in front of the sick person and see his disease reflected in it; then I can cure the disease." A Northern Arapaho medicine bundle collected in 1923 for the Museum of the American Indian contained an elaborately embellished mirror board, 9 1/2 in. by 3 1/2 in., among other objects.[66]

Sometimes the flash of the mirror was evoked even in the absence of the object itself. The Dakota Sioux did beadwork in a series of patterns they called "looking-glass patterns" or "reflected patterns." These complex designs suggested "confusion of eye-movements when looking at them." Wissler posited "that the effect of such a combination of lines and areas upon the observer" led to the name of this pattern, which was "certainly somewhat analogous to the flashing of a mirror in the face."[67] These looking-glass patterns emphasize the glittery deflection of light that mirrors can produce to disorient human vision, one of the key ways in which mirrors both were put to practical use (momentarily blinding an enemy) and helped make meaning (in ritual uses) in Native American societies.

The introduction and proliferation of European-made mirrors into Native American and African American societies connected men and women to an earlier embrace of using reflective and glittery objects to harness sacred power, giving practitioners easier access to the tools they needed to accomplish their goals. Mirrors, as we have seen, also offered people extended opportunities to engage with an accurate image of themselves and to construct a self-identity that relied at least partially on that image, a quintessentially modern practice.[68] Men and women of African, Native, and European descent embraced the mirror's ability to make self-knowledge available to them. But in none of these cultures did this embrace completely eradicate the mirror's power, for it continued to be revered and ritualized across nineteenth-century North America.

Epilogue

Immediately following the Civil War, Mary Ames, a northern white woman, headed to South Carolina to teach for the Freedman's Bureau on Edisto Island. To take this job, she had to leave behind many of the comforts of home: she and her female traveling companion brought with them only "a chair, a plate, knife, fork and spoon; cup and saucer, blanket, sheets and pillowcases, and sacking for a bed of hay or straw to be found wherever we should be situated, and we added some crackers, tea, and a teapot." Not long after their arrival on Edisto Island, the teachers received boxes of clothing to distribute to local African Americans. The islanders would have to "pay for [the clothing] with vegetables, eggs, chickens, or whatever they can bring in exchange." One man came looking not for clothes but for a gun. What did he think might entice these women to provision him with a weapon (if they even had one)? In addition to potatoes, he offered them a "cracked looking-glass." The teachers did not fulfill his request for a weapon, but they did trade some articles of clothing for the looking glass and potatoes.[1] This man clearly assumed he knew enough about white women to know that this would be a desirable item to them (which was borne out by the women's reaction), one that might even garner him a weapon with which he could hunt and defend himself and his family.

Although the exchange that took place on Edisto Island revealed the complex patterns of ownership of and meanings associated with mirrors in the nineteenth century, in popular culture whites continued to assert that African American ownership and use of mirrors was something to be mocked.[2] In 1866 the "Humors of the Day" feature in *Harper's Weekly* included "Selections from the Constitution of the New Freedmen's Bureau." Undoubtedly written by a white author, the "Constitution" issued these demands:

1. Every freedman shall have a bureau for himself, with a looking-glass on the top, if he wants it.
2. Every freedman shall have a secretary.
3. Every freedboy or freedgirl shall have a wardrobe.
4. Every freedchild shall have whatever it cries for.
5. White people, whether free or not, must behave themselves.
6. All persons of every color, except red, must vote.[3]

The "Constitution of the New Freedmen's Bureau" tells us nothing about what African Americans may really have wanted or needed as they made the transition from slavery to freedom, much less so than the exchange that took place on Edisto Island at war's end. What it does reveal, however, are whites' prejudices against and assumptions about African Americans. The Freedmen's Bureau was established in 1865 by an Act of Congress to provide material, educational, and legal aid to recently freed African Americans in the post–Civil War South. The *Harper's Weekly* feature ridiculed the very existence of the Freedmen's Bureau by turning the "Bureau" in the agency's title into its other possible meaning: a chest of drawers. The mockery continued by placing a looking glass on the top of this fictional chest of drawers—a common placement at the time—and indicating that a freedman would be offered the glass "if he wants it." White Americans understood mirrors to be a necessary tool. Of course, whites would be assumed to have wanted a mirror if offered to them. But African Americans might not want this object, the *Harper's Weekly* feature insinuated, because they might not recognize its value. This was the fiction that whites had been telling themselves for centuries. Whites continued to assert that they were the only true inheritors of the Enlightenment-era confidence in vision as the noblest sense and therefore the only ones who could use mirrors for their highest purposes—as a rational tool of self-discovery and extender of human vision. Whites believed that those they had deemed inferior to themselves—in this case African Americans—did not have access to this highest-order vision and therefore might misunderstand the mirror's significance and even reject it when offered to them.

After the Civil War, as white men feared that their mastery over African Americans would slip away, perhaps they retained some comfort in their superiority over white women—like those teachers on Edisto Island—who continued to rely on the looking glass and to see themselves through the male gaze it provided them. Yet as the rumblings from Seneca Falls in 1848 had intimated, some white women were increasingly prepared to challenge these roles that men had allotted to them. A group of some three hundred women and men met in Seneca Falls, New York, to seek full rights for women in American society and had written the Declaration of Sentiments to express their grievances, based on the Declaration of Independence. An image from 1871 would have confirmed male fears about women's demands to be full participants in the life of the nation. In it, a woman advocate of equal rights practices a speech while a man stands beside her, doing laundry, a reversal of traditionally constructed gender roles among whites. The woman stands before a looking glass—not to judge the acceptability of her personal front through the male gaze, but to imagine herself standing before a group of listeners as she demands full equality with men (fig. E.1).

As late nineteenth-century Native peoples' status as distinct nations were threatened by the Civil War–era massacres, continued removal, and ultimately the reservation system, Native peoples continued to seek the power of the sacred to protect them from white expansion, culminating in the war dances of the late nineteenth century. One war dance shirt has three paper looking glasses sown into the center of the chest, in a straight line descending just below the collar, on the front and back of the shirt (fig. E.2). Some practitioners of the Ghost Dance believed that these Ghost Shirts had the power to protect their wearers from the bullets they would encounter in battle.[4] The mirrors,

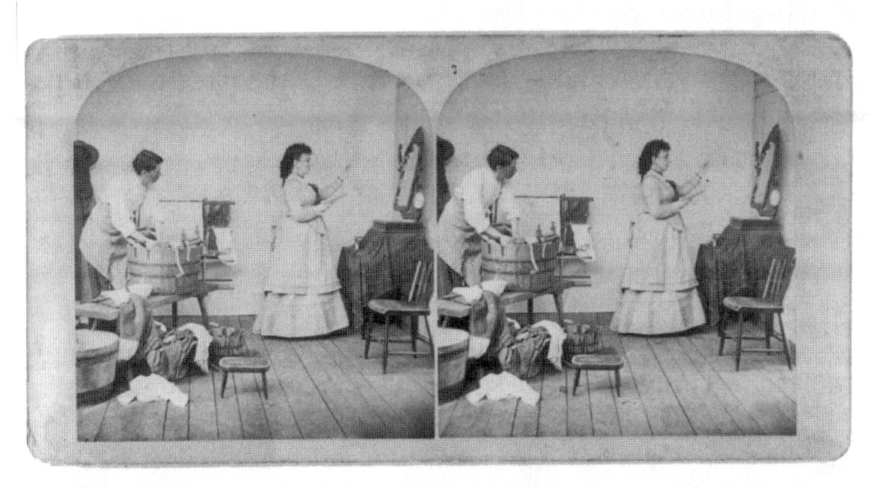

Figure E.1. Women's Rights: The Rehearsal, ca. 1871. Courtesy of Library of Congress Prints and Photographs Division.

Figure E.2. War shirt, Blackfeet (uncertain). Courtesy of Penn Museum, image no. 229029.

with their deflective potential, may have been seen as an added element of protection. Highly decorated hand mirrors, framed and decorated by Plains Indians, also played a role in the Ghost Dance.[5]

For three centuries, white men had built a racial and gender hierarchy that placed them in a position of power over white women, Native Americans, and African Americans in North America. The looking glass was an important material object through which they believed that they displayed their mastery over the world and in which they could see the inferiority of those they desired to subjugate. Yet they were never able to completely control the range of meanings and practices that different groups associated with looking glasses. By the late nineteenth century, white women, Native Americans, and African Americans claimed ownership of and the right to determine the meaning of mirrors. Even though none of them would achieve the right to self-determination that they desired during this period, they successfully began to challenge the ability of American gender and racial hierarchies to define the meaning of the mirror, an object intimately linked to identity formation, for them and to claim, as well, the right to define their own identities for themselves.

Acknowledgments

It takes a village, so the saying goes, to raise a child. It also takes one to write a book, especially when also raising a child, the two activities that have defined my days over the past several years. I am thankful to everyone who has encouraged and supported me in ways large and small during this time.

This project began as a dissertation written in the History Department at the University of South Carolina. The faculty, staff, and students there made it an exceptional place to be a graduate student. Mark Smith, who directed my PhD dissertation; Jessica Kross, who directed my MA thesis; and Katherine Grier, who guided my work in museums and material culture, were exemplary professional role models. As a scholar, teacher, and public historian, I am deeply in their debt. I am very thankful as well to the wider circle of faculty who have mentored me, especially Kathryn Edwards, Larry Glickman, Ann Johnson, Connie Schulz, and Lynn Weber. My graduate school colleagues included Kate Jones, Francesca Fair, Aaron Marrs, Kathy Hilliard, Mike Reynolds, Eric Plaag, Kevin Dawson, Sara Burrows, Jay Richardson, Melissa Jane Taylor, Linda Ziegenbein, and David Prior. I am grateful for their generosity and for their friendship. While in graduate school, I was extraordinarily lucky to have been befriended by the Joyner family—Charles, Jean, Wesley, and Hannah. Their love and encouragement have meant so much to me. Charles, who died before the publication of this volume, was a model for many in my generation of what a generous and engaged academic life could look like, and I am grateful for that example and for his friendship.

I am indebted to the librarians, archivists, and curators at the South Caroliniana Library, Thomas Cooper Library, Stevens Memorial Museum, Museum of Early Southern Decorative Arts, Winterthur Museum, Pitt Rivers Museum, Vedder Memorial Library, Museum of the American Indian, Canadian Museum of Civilization, and the Field Museum of Natural History in Chicago. Special thanks goes to Teodora Durbin at the IUPUI library who has provided me with access to sources over the past five years. I also want to thank graduate students Caitlyn Stypa, Nancy Brown, and Rebecca Denne, who assisted me with this project.

This project has benefited from formal and informal feedback in a wide variety of settings. I experimented with my early ideas about the meanings of mirrors in early America at the meetings of the American Studies Association, the American Historical Association, and the Organization of American Historians. In more recent years I benefited enormously from conversations with colleagues and those in attendance at the Seventeenth-Century Warfare, Diplomacy, and Society in the American Northeast conference held at the Mashantucket Pequot Museum and Research Center and the Seminar on Colonial Objects and Social Identity held at the National Museum of Denmark.

It has been a privilege to publish this book with Johns Hopkins University Press. I am very grateful to the staff at the press, especially Elizabeth Demers and Meagan Szekely, as well as to Joe Abbott, for their work on this project. I am so grateful for Elizabeth's enthusiastic support from the very beginning, Meagan's prompt and friendly replies to countless emails from me, and Joe's attention to detail as the project neared completion.

I have been fortunate to work in two collegial history departments—first at the University of Wisconsin at Whitewater and now at Indiana University–Purdue University Indianapolis. My work has been made much more enjoyable by friends and colleagues in both of these places. I am thankful to Nikki Mandell and Tony Gulig at UWW, who served as role models as I began my academic career. My writing group, consisting of Ellie Schemenauer and Deborah Wilk, was a highlight of my time there. I am especially grateful to Mark Boulton for his friendship and support and for always knowing how to make me laugh. At IUPUI, I am grateful for the friendship and guidance I have received from Danna Kostroun, Modupe Labode, Liz Monroe, Nancy Robertson, Stephanie Rowe, Phil Scarpino, and Elee Wood. Jen Guiliano deserves special thanks for the critical eye she brought to my work in its final stages, for navigating the tenure process with me, and for her friendship. I am also thankful to my public history tribe. Marla Miller finessed the manuscript and changed it significantly for the better. Cherstin Lyon and Betsy Nix supported me as I juggled our project with finishing this book. John Dichtl is a fantastic friend and colleague. One of the best things about being at IUPUI was getting to know and work with him before he moved on to head the American Association of State and Local History.

It is hard to know how to thank the people closest to my heart. I have known Chandra Yoder and Sophie Stanes since our days at the University of Warwick, and although we don't see each other as often as we would like, they are two of the people I can't imagine what my life would be like without. Kathy Hilliard and Linda Ziegenbein have been my friends since our grad school days at USC. Kathy has been a step or two ahead of me throughout our careers and has always generously shared her time, expertise, and advice. Our friendship was forged around a seminar table in Gambrell Hall and as neighbors on Deerwood Street, and I am so grateful for it. Linda was a step or two ahead of me on the path to motherhood (two steps precisely: Malie and Lila). She gave me the courage to believe that I could finish this book and be a mother, making possible the greatest joy of my life. Linda has also read more drafts of my writing than anyone else and has tirelessly supported me as I navigated the academy and motherhood. Someday we are going to enjoy museum visits again when we do not have to spend all of our time in the kid-friendly areas! Margie Ferguson and I have walked countless laps on the NIFS track at IUPUI, and our conversations have made my struggles seem much more manageable. I am so grateful for her daily presence in my life. I would not be here without this far-flung community of women who support and love one another so fiercely.

My parents, Larry and Pat Shrum, have been my biggest supporters throughout my life, and I am grateful for their love and encouragement. This book is dedicated to my son, Brady, who brightens my life every day without fail. A few years ago, Brady and I welcomed Craig Brooks into our family. This book, as Craig would say, has been "just one of those things." His perspective helps keep me grounded, and I am so thankful to him for his love and support.

Notes

Introduction

1. *Looking glass* was the common term in early America for a glass mirror. Throughout this work I use *looking glass* and *mirror* interchangeably to describe this object. A discussion of the distinctions between these terms can be found in chapter 1.

2. Zebulon Montgomery Pike, *The Expeditions of Zebulon Montgomery Pike*, ed. Elliot Coues (New York: Francis P. Harper, 1895), 1:83–90; Mary Lethert Wingerd, *North Country: The Making of Minnesota* (Minneapolis: University of Minnesota Press, 2010), 76–77. In 1805 the first governor of the Louisiana Territory commissioned Zebulon Pike to lead an expedition that would establish an American presence in the northern part of the territory and challenge British primacy in the fur trade. Departing from Fort Bellefontaine, just north of St. Louis, Pike's party traveled upriver. After six weeks, they reached what would become Fort Snelling and negotiated a treaty with local Mdewakanton Dakota, who received gifts Pike valued at $250, a promise of $2,000, and retained use rights on the land.

3. George Lakoff and Mark Johnson, *Metaphors We Live By* (Chicago: University of Chicago Press, 1980), 36–37.

4. George Catlin, *Illustrations of the Manners, Customs, and Condition of the North American Indians* (London: Henry G. Bohn, 1848), 2:221–22.

5. *Mirror self* is the term I use throughout this volume to describe the reflection of oneself made visible in a mirror.

6. George P. Rawick, ed., *The American Slave: A Composite Autobiography*, vol. 18, *Unwritten History of Slavery* (Westport, CT: Greenwood, 1972), 263–64.

7. Jacob D. Green, *Narrative of the Life of J. D. Green, a Runaway Slave, from Kentucky* . . . (Huddersfield, UK: Henry Fielding, Pack Horse Yard, 1864), 11, http://docsouth.unc.edu/neh/greenjd/greenjd.html.

8. The historical study of whiteness is rooted in the position, advocated here and articulated by critical race theorists, that "race and races are products of social thought and relations. Not objective, inherent, or fixed, they correspond to no biological or genetic reality; rather, races are categories that society invents, manipulates, or retires when convenient. People with common origins share certain physical traits, of course, such as skin color, physique, and hair texture. But these constitute only an extremely small portion of their genetic endowment, are dwarfed by that which we have in common, and have little or nothing to do with distinctly human, higher-order traits, such as personality, intelligence, and moral behavior." Richard Delgado and Jean Stefancic, *Critical Race Theory: An Introduction*, 2nd ed. (New York: New York University Press, 2012), 8–9. For a discussion of the cultural

construction of race in North America see Winthrop Jordan, *White over Black: American Attitudes toward the Negro, 1550–1812* (Chapel Hill: University of North Carolina Press, 1974); and Mia Bay, *The White Image in the Black Mind: African-American Ideas about White People, 1830–1925* (New York: Oxford University Press, 2000). Two extensive reviews of whiteness scholarship are Peter Kolchin, "Whiteness Studies: The New History of Race in America," *Journal of American History* 89, no. 1 (2002): 154–73; and Daniel Wickberg, "Heterosexual White Male: Some Recent Inversions in American Cultural History," *Journal of American History* 92, no. 1 (2005): 136–57. A key early work on the subject is Barbara Fields, "Slavery, Race and Ideology in the United States of America," *New Left Review* 181 (May-June 1990): 95–118.

9. By the late twentieth century, an American company had invented a "true mirror" to overcome the latter distortion in the mirror's reflection (based on what the company called a "crude" attempt first patented in 1887). The true mirror "is made from putting two mirrors at a right angle, such that the two mirror images bounce off of each other." Thus what one sees in this mirror is the same as what an observer sees: if you extend your right hand in a true mirror, your mirror self also extends its right hand. See "More Information—How It Works," True Mirror, www.truemirror.com/Moredata.asp. Recent studies have suggested that the difference between our mirror selves and the selves we present to other observers can be profound because of this reversal, but early Americans seemed altogether unconcerned about the distinction or, at least, did not leave any trace of such a concern in the evidence now available. See "Desperately Seeking Symmetry," Radio Lab podcast (produced by WNYC), season 9, episode 5, www.radiolab.org/story/122382-desperately-seeking-symmetry/.

10. For the psychological perspective on the importance of mirrors in shaping human identity, see Jacques Lacan, "The Mirror Stage as Formative of the Function of the I as Revealed in Psychoanalytic Experience," in *Écrits: A Selection*, trans. Alan Sheridan (New York: Norton, 1977), 1–7; and James Elkins, *The Object Stares Back: On the Nature of Seeing* (San Diego: Harvest, 1996), 69–75. Although neuroscientists are not convinced by Lacan, many still take mirrors seriously. See Julian Paul Keenan, with Gordon G. Gallup Jr. and Dean Falk, *The Face in the Mirror: The Search for the Origins of Consciousness* (New York: Ecco, 2003); Gordon G. Gallup Jr., "Chimpanzees: Self-Recognition," *Science* 167, no. 3914 (1970): 86–87; and V. S. Ramachandran and E. L. Altschuler, "The Use of Visual Feedback, in Particular Mirror Visual Feedback, in Restoring Brain Function," *Brain* 132, no. 7 (2009): 1693–1710. For perspectives from literature and art history see Jenijoy La Belle, *Herself Beheld: The Literature of the Looking Glass* (Ithaca, NY: Cornell University Press, 1989); and Jonathan Miller, *On Reflection* (London: National Gallery, 1998). For a historical work heavily indebted to Lacan that explores religious, philosophical, and psychological meanings associated with mirrors (almost exclusively in France) as revealed in a wide range of literary and artistic sources, see Sabine Melchior-Bonnet, *The Mirror: A History*, trans. Katharine H. Jewett (New York: Routledge, 2001); and Wendy Doniger's review of *The Mirror: A History*, "Lacan's Ghost," *London Review of Books* 24, no. 1 (2002): 7–8. An additional work that looks at human interactions with mirrors throughout history is Mark Pendergrast, *Mirror, Mirror: A History of the Human Love Affair with Reflection* (New York: Basic Books, 2003). One important exception to the lack of historical work on the role of mirrors in shaping early American identities is historian Ann Smart Martin, *Buying into the World of Goods: Early Consumers in Backcountry Virginia* (Baltimore: Johns Hopkins University Press, 2008). Martin explores why ordinary men and women in eighteenth-century Virginia valued consumer goods, including a small mirror purchased by Suckey, an enslaved woman of African descent (186–93).

11. Henry Glassie, *Material Culture* (Bloomington: Indiana University Press, 1999), 41; Leora Auslander, "Beyond Words," *American Historical Review* 110, no. 4 (2005): 1015; Thorstein Veblen introduced the idea of emulation in *The Theory of the Leisure Class* (1899; repr., Mineola, NY: Dover Thrift, 1994), 15–22. For historians' use of the theory of emulation see Neil McKendrick, John Brewer, and J. H. Plumb, *The Birth of a Consumer Society: The Commercialization of Eighteenth-Century England* (Bloomington: Indiana University Press, 1982); and Richard Bushman, *The Refinement of America: Persons, Houses, Cities* (New York: Vintage, 1992).

12. Arjun Appadurai, ed., *The Social Life of Things: Commodities in Cultural Perspective* (Cambridge: Cambridge University Press, 1988); T. H. Breen, "The Meanings of Things: Interpreting the Consumer Economy in the Eighteenth Century," in *Consumption and the World of Goods*, ed. John Brewer and Roy Porter (New York: Routledge, 1997), 250; Amanda Vickery, "Women and the World of Goods: A Lancashire Consumer and Her Possessions, 1751–81," in *Consumption and the World of Goods*, ed. John Brewer and Roy Porter (New York: Routledge, 1997), 277; Paul Clemens, "The Consumer Culture of the Middle Atlantic, 1760–1820," *William and Mary Quarterly*, 3rd ser., 62, no. 4 (2005): 577–624.

13. Mihaly Csikszentmihalyi and Eugene Rochberg-Halton, *The Meaning of Things: Domestic Symbols and the Self* (Cambridge: Cambridge University Press, 1981), 13–19.

14. Carolyn Gilman, *Where Two Worlds Meet: The Great Lakes Fur Trade* (St. Paul: Minnesota Historical Society Press, 1982), 104.

15. As Chris Jenks observes, "visual culture" originally referred to "painting, sculpture, design and architecture," but in recent years it has broadened considerably. Visual culture now includes studies of what John Davis terms "practices of vision," which seek to understand "the historically and socially shaped character of vision." It is now well established, as Martin Jay has argued, that while there are "certain fairly fundamental characteristics" of sight that "no amount of cultural mediation can radically alter," vision is always historically situated and contingent. As the field of visual culture studies has expanded, so, too, has the range of practices, images, and objects that scholars have considered in seeking to better understand the historical construction of vision. My work seeks to situate vision historically in North America as it was made manifest to early Americans through their encounters with the looking glass. See Chris Jenks, "The Centrality of the Eye in Western Culture: An Introduction," in *Visual Culture*, ed. Chris Jenks (London: Routledge, 1995), 16; John Davis, "The End of the American Century: Current Scholarship on the Art of the United States," *Art Bulletin* 85, no. 3 (2003): 560; and Martin Jay, *Downcast Eyes: The Denigration of Vision in Twentieth-Century French Thought* (Berkeley: University of California Press, 1994), 5.

16. Depending on the skill of the artist and the style of painting, early Americans with access to portraiture did have an additional aid in developing this deep knowledge about the appearance of the self. See David Jaffee, "One of the Primitive Sort: Portrait Makers of the Rural North, 1760–1860," in *The Countryside in the Age of Capitalist Transformation: Essays in the Social History of Rural America*, ed. Steven Hahn and Jonathan Prude (Chapel Hill: University of North Carolina Press, 1985), 103–38; Caroline F. Sloat, ed., *Meet Your Neighbors: New England Portraits, Painters, and Society, 1790–1850* (Sturbridge, MA: Old Sturbridge Village, 1992); and David Jaffee, *A New Nation of Goods: The Material Culture of Early America* (Philadelphia: University of Pennsylvania Press, 2010), 218–73.

17. Clifford Geertz, "'From the Native's Point of View': On the Nature of Anthropological Understanding," *Bulletin of the American Academy of Arts and Science* 28, no. 1 (1974): 31.

18. On the presumed role played by mirrors in the creation of the modern sense of self during the Renaissance, see Alan MacFarlane and Gerry Martin, *Glass: A World History*

(Chicago: University of Chicago Press, 2002), 70–72; and Benjamin Goldberg, *The Mirror and Man* (Charlottesville: University Press of Virginia, 1985), 160. Although not focused on the role of mirrors in shaping human identity, Dora Thornton's *The Scholar in His Study: Ownership and Experience in Renaissance Italy* (New Haven, CT: Yale University Press, 1997), provides rich descriptions of Renaissance mirrors.

19. Roy Porter, introduction to *Rewriting the Self: Histories from the Renaissance to the Present*, ed. Roy Porter (London: Routledge, 1997), 1–4. Porter's volume makes significant progress in dismantling this fiction.

20. On the rising importance of the individual during the Renaissance see Jacob Burckhardt, *The Civilization of the Renaissance in Italy* (1860; repr., New York: Modern Library, 2002); Stephen Greenblatt, *Renaissance Self-Fashioning from More to Shakespeare* (Chicago: University of Chicago Press, 1980); and Charles Taylor, *Sources of the Self: The Making of the Modern Identity* (Cambridge, MA: Harvard University Press, 1989).

21. Debora Shuger, "The 'I' of the Beholder: Renaissance Mirrors and the Reflexive Mind," in *Renaissance Culture and the Everyday*, ed. Patricia Fumerton and Simon Hunt (Philadelphia: University of Pennsylvania Press, 1999), 35, 37. Shuger does conclude that the individual sense of self had developed by the late seventeenth century. Looking specifically at England, Dror Wahrman argues in *Making of the Modern Self: Identity and Culture in Eighteenth-Century England* (New Haven, CT: Yale University Press, 2006) that the "core of selfhood characterized by psychological depth, or interiority, which is the bedrock of unique, expressive individual identity," did not emerge in England until the end of the eighteenth century (xi). Wahrman does not consider the role of mirrors in this process. I do not aim here to weigh in on when the modern sense of self emerged in Western Europe or England, but I am sympathetic to the counterarguments offered by Shuger and Wahrman because the emergence of the modern self, at least in relation to people's interactions with mirrors, certainly developed unevenly over several centuries in North America.

22. Michael McKeon, "Recent Studies in the Restoration and Eighteenth Century," *Studies in English Literature, 1500–1900* 45, no. 3 (2005): 714.

CHAPTER 1: The Evolving Technology of the Looking Glass

1. Duke of Saint-Simon, *Memoirs of Louis XIV and the Regency*, trans. Bayle St. John (Washington: M. Walter Dunne, 1901), 1:166. The English referred to all grain as "corn." What the translator likely meant here was wheat.

2. The Venetian monopoly and the attempts by other European nations to undercut it is a tale full of intrigue recounted in Sabine Melchior-Bonnet, *The Mirror: A History*, trans. Katharine H. Jewett (New York: Routledge, 2001), 30–46; and Benjamin Goldberg, *The Mirror and Man* (Charlottesville: University Press of Virginia, 1985), 163–70.

3. Melchior-Bonnet, *The Mirror*, 29; Lorna Weatherill, *Consumer Behaviour and Material Culture in Britain, 1660–1760*, 2nd ed. (New York: Routledge, 1996), 3, 26–27. See also Eleanor S. Godfrey, *The Development of English Glassmaking, 1560–1640* (Chapel Hill: University of North Carolina Press, 1975), 235–41; and Geoffrey Wills, *English Looking-Glasses: A Study of the Glass, Frames, and Makers, 1670–1820* (New York: A. S. Barnes, 1965), 41–44.

4. See W. E. Minchinton, *The British Tinplate Industry: A History* (Oxford: Clarendon, 1957), 1.

5. F. W. Gibbs, "The Rise of the Tinplate Industry. I. The Tinplate Workers," *Annals of Science* 6, no. 4 (1950): 398; and Minchinton, *The British Tinplate Industry*, 4, 11.

6. See Wills, *English Looking Glasses*, 41, 143–44; and Geoffrey Wills, "From Polished Metal to Looking-Glass," *Country Life*, Oct. 23, 1958, 939. A remaining example of a specu-

lum can be seen in the collection of the Cotehele House, Cornwall, England. This irregularly shaped mirror is 10 in. × 12 in. × 0.25 in. (deep) and weighs slightly more than six pounds. "Cotehele Speculum," *National Trust*, www.nationaltrustcollections.org.uk/object/347879.

7. See Ingeborg Krueger, "Glass-Mirrors in Medieval Times," in *Annales du 12e Congrès de l'association international pour l'histoire du verre* (Amsterdam: Association internationale de l'histoire du verre, 1993), 329–31. For an insightful discussion of a group of medieval ivory-carved mirror frames see Susan L. Smith, "The Gothic Mirror and the Female Gaze," in *Saints, Sinners, and Sisters: Gender and Northern Art in Medieval and Early Modern Europe*, ed. Jane L. Carroll and Alison G. Stewart (Burlington, VT: Ashgate, 2003), 73–93.

8. One piece of the mirror glass had a maximum diameter of 23 mm and the other 25 mm. See Martin Biddle, *Object and Economy in Medieval Winchester* (Oxford: Clarendon, 1990), 2:653–58. Biddle doubts whether the "two tuns" should be read literally because given the weight of the Winchester mirror, "two tuns" would amount to 134,400 mirrors. He asks whether, perhaps, "two tuns" indicated "a barrel or cask used as a container, unrelated . . . to the ton weight?" (2:654).

9. Bruno Schweig, *Mirrors: A Guide to the Manufacture of Mirrors and Reflecting Surfaces* (London: Pelham, 1973), 20; Alexa Sand, *Vision, Devotion, and Self-Representation in Late Medieval Art* (New York: Cambridge University Press, 2014), 64, 313n127; "About Medieval Glass," Corning Museum of Glass, www.cmog.org/article/about-medieval-glass.

10. John E. Crowley, *The Invention of Comfort: Sensibilities and Design in Early Modern Britain and Early America* (Baltimore: Johns Hopkins University Press, 2001), 123; Godfrey, *Development of English Glassmaking*, 235. Crowley notes that Chaucer's (ca. 1343–1400) "references to glass mirrors usually implied dullness" (317n20). Further complicating an assessment of these early metal and glass mirrors is how few have survived. This author has never seen one, but even if one could be viewed, the deterioration caused over the centuries would make it difficult to judge the original degree of reflectivity.

11. Peter Radisson, *Voyages of Peter Esprit Radisson* (London: Publications of the Prince Society, 1858), www.gutenberg.org/ebooks/6913.

12. Krueger, "Glass-Mirrors in Medieval Times," 322–25. For more on Venetian glassmaking see Goldberg, *The Mirror and Man*, 139–42; and W. Patrick McCray, "Glassmaking in Renaissance Italy: The Innovation of Venetian Cristallo," *Journal of the Minerals, Metals and Materials Society* 50, no. 5 (1998): 14.

13. Ivor Noël Hume, *A Guide to Artifacts of Colonial America* (New York: Knopf, 1985), 233–35; Goldberg, *The Mirror and Man*, 142; Karl G. Roenke, *Flat Glass: Its Use as a Dating Tool for Nineteenth Century Archaeological Sites in the Pacific Northwest and Elsewhere* (Moscow, ID: Department of Sociology/Anthropology, University of Idaho, 1978), 6–7.

14. Tony Spawforth, *Versailles: A Biography of a Palace* (New York: St. Martin's, 2008), 33; Roenke, *Flat Glass*, 9; Helen Comstock, *The Looking Glass in America, 1700–1825* (New York: Viking, 1968), 16.

15. Andrew Ure, *A Dictionary of Arts, Manufactures, and Mines* (New York: D. Appleton, 1858), 1:920.

16. John Doggett and Company Letterbook, 1825–1829, Joseph Downs Manuscript Collection, Winterthur Museum, Garden and Library, quoted in David L. Barquist, *American Tables and Looking Glasses in the Mabel Brady Garvan and Other Collections at Yale University* (New Haven, CT: Yale University Art Gallery, 1992), 295; Kenneth M. Wilson, "Plate Glass in America: A Brief History," *Journal of Glass Studies* 43 (2001): 145.

17. Schweig, *Mirrors*, 23–24; Wills, *English Looking-Glasses*, 64. Fortunately for the student of material culture, this coating also ages differently than did its predecessor. The

tin-mercury amalgam "tends to go grey, in patches often circular in shape, and glittering where tiny granules of metal have isolated themselves," but the more modern silvering "will tarnish completely where the air has reached it" and does not break down into granules (Wills, *English Looking-Glasses*, 64). (See fig. 1.4 for an example of the small gray patches that appear in quicksilvered looking glasses.)

18. Mary Ellen Hayward, "The Elliotts of Philadelphia: Emphasis on the Looking Glass Trade, 1755–1810" (master's thesis, University of Delaware, 1971), 124–25; Faverie Advertisement, *Virginia Patriot and Richmond Daily Mercantile Advertiser*, Feb. 5, 1818, 3–5; Index of Early Southern Artists and Artisans, Museum of Early Southern Decorative Arts, Winston-Salem, North Carolina.

19. The *Oxford English Dictionary* lists the earliest usage of *mirror* in English as ca. 1250 and the earliest usage of *looking glass* as 1526. "Mirror, n.," and "Looking glass, n.," *Oxford English Dictionary*, www.oed.com. That the word *looking glass* does not emerge until after accurately reflective looking glasses began to be widely made suggests that there was not enough difference between the earlier metal and glass mirrors for there to be a need to differentiate them linguistically.

20. John Pynchon, "Account Books of Major John Pynchon, Springfield, Massachusetts," vol. 1 (1651–55), vol. 2 (1657–66) [microfilm reel 1] (Springfield: Connecticut Valley Historical Museum, 1957); Radisson, *Voyages of Peter Esprit Radisson*, www.gutenberg.org/ebooks/6913. See chapter 2 for a discussion of the Pynchon records.

21. Giovanni da Verrazano, "The Written Record of the Voyage of 1524 of Giovanni da Verrazano," adapted from *The Voyages of Giovanni da Verrazano, 1524–1528*, ed. Lawrence C. Wroth, trans. Susan Tarrow (New Haven, CT: Yale University Press, 1970), 133–43, www.columbia.edu/~lmg21/ash3002y/earlyac99/documents/verrazan.htm; Fray Antonio de la Ascensión, "Brief Report of the Discovery in the South Sea," in *Spanish Exploration in the Southwest, 1542–1706*, ed. Herbert E. Bolton (New York: C. Scribner's Sons, 1916), 125, www.americanjourneys.org/aj-003/.

22. Marc Lescarbot, *History of New France: With an English Translation, Notes and Appendices* (New York: Greenwood, 1968), 1:61; Daniel Francis and Toby Morantz, *Partners in Furs: A History of the Fur Trade in Eastern James Bay, 1600–1870* (Kingston, ON: McGill-Queen's University Press, 1983), 16.

23. James A. Bruseth and Toni S. Turner, *From a Watery Grave: The Discovery and Excavation of La Salle's Shipwreck, "La Belle"* (College Station: Texas A&M University Press, 2005), 90–91. In 1995 archaeologists began to excavate *La Belle*, which sank during a storm in Matagorda Bay along the Texas coastline, eventually uncovering more than one million artifacts that provide a rare glimpse at what the French brought to establish a colony in North America at the end of the seventeenth century.

24. Conrad Hilberry, ed., *The Poems of John Collop* (Madison: University of Wisconsin Press, 1962), 93–94.

25. Barquist, *American Tables and Looking Glasses*, 294. Barquist does note that *looking glass* and *mirror* may have been occasionally used as synonyms and that Samuel Johnson treated them as such in his 1755 *Dictionary of the English Language*. But the "common name" was *looking glass* or *glass*. Sometime in the late nineteenth century—Barquist suggests after 1860—the term *mirror* became the standard term for this object (294).

26. Probate Inventory of William Parsell, Newtown, Queens, New York, Inventories of Estates, New York City and Vicinity, 1717–1844, New-York Historical Society. See also the Last Will and Testament of William Parcell [*sic*] (December 22, 1724), in *Collections of the New-York Historical Society for the Year 1902* (New York: Printed for the Society, 1903), 57–58.

27. Although it is commonly assumed that the combination of candle and mirror increased the amount of light in a room, Nancy Carlisle argues that "modern light meter tests show that a candle backed by a looking glass puts out an insignificant amount of light over a candle alone." Nancy Camilla Carlisle, "A Reflection of the Times: The Looking Glass in Eighteenth-Century America" (master's thesis, University of Delaware, 1983), 54–55.

28. Elisabeth Donaghy Garrett, *At Home: The American Family, 1750–1870* (New York: Henry J. Abrams, 1990), 153.

29. An expansive visual record of the eighteenth- and early nineteenth-century chimney and pier glasses can be found in Barquist, *American Tables and Looking Glasses*, 294–341.

30. Thomas Sheraton, *The Cabinet Dictionary* (London: W. Smith, 1803), 201, 236, https://books.google.com/books/about/The_cabinet_dictionary_To_which_is_added.html?id=ov5bAAAAQAAJ.

31. Wills, *English Looking-Glasses*, 35–36, 128; Sheraton, *The Cabinet Dictionary*, 271; Goldberg, *The Mirror and Man*, 174. Sheraton also indicated that any convex or concave piece of glass was known by the term *mirror* (271).

32. Elisabeth Donaghy Garrett, "Looking Glasses in America: 1700–1850," in David L. Barquist, *American Tables and Looking Glasses in the Mabel Brady Garvan and Other Collections at Yale University* (New Haven, CT: Yale University Press, 1992), 30–31.

CHAPTER 2: First Glimpses

1. *The Sovereignty and Goodness of God by Mary Rowlandson with Related Documents*, ed. Neal Salisbury (Boston: Bedford/St. Martin's, 1997), 5, 25–31, 34, 96; Laurel Thatcher Ulrich, *Good Wives: Image and Reality in the Lives of Women in Northern New England, 1650–1750* (New York: Vintage, 1991), 227.

2. Giovanni da Verrazano, "The Written Record of the Voyage of 1524 of Giovanni da Verrazano," adapted from *The Voyages of Giovanni da Verrazano, 1524–1528*, ed. Lawrence C. Wroth, trans. Susan Tarrow (New Haven, CT: Yale University Press, 1970), 133–43, www.columbia.edu/~lmg21/ash3002y/earlyac99/documents/verrazan.htm. On Native peoples' initial reactions to Europeans see Evan Haefeli, "On First Contact and Apotheosis: Manitou and Men in North America," *Ethnohistory* 54, no. 3 (2007): 407–43; and Bruce G. Trigger, "Early Native North American Responses to European Contact: Romantic versus Rationalistic Interpretations," *Journal of American History* 77, no. 4 (1991): 1195–1215.

3. See Arthur J. Ray, "Indians as Consumers in the Eighteenth Century," in *Old Trails and New Directions: Papers of the Third North American Fur Trade Conference*, ed. Carol M. Judd and Arthur J. Ray (Toronto: University of Toronto Press, 1980), 255–71; Daniel Francis and Toby Morantz, *Partners in Furs: A History of the Fur Trade in Eastern James Bay, 1600–1870* (Kingston, ON: McGill-Queen's University Press, 1983), 61; and K. G. Davies, *Letters from Hudson Bay, 1703–40* (London: Hudson's Bay Record Society, 1965), 27, 278–84.

4. "A Letter of William Bradford and Isaac Allerton, 1623," *American Historical Review* 8, no. 2 (1903): 295; John Lederer, *The Discoveries of John Lederer . . .*, ed. William P. Cumming (Charlottesville: University Press of Virginia, 1958), 27; Bruce G. Trigger and William R. Swagerty, "Entertaining Strangers: North America in the Sixteenth Century," in *The Cambridge History of the Native Peoples of the Americas*, ed. Bruce G. Trigger and Wilcomb E. Washburn, vol. 1, pt. 1, *North America* (Cambridge: Cambridge University Press, 1996), 377–78.

5. William Pynchon, "Bill of William Pynchon to John Winthrop Jr.," *Winthrop Papers*, vol. 3, *1631–1637* (Boston: Massachusetts Historical Society, 1943), 238; Ruth A. McIntyre, *William Pynchon: Merchant and Colonizer, 1590–1662* (Springfield, MA: Connecticut Valley

Historical Museum, 1961), 9–13; Walter W. Woodward, *Prospero's America: John Winthrop, Jr., Alchemy, and the Creation of New England Culture, 1606–1676* (Chapel Hill: University of North Carolina Press, 2010), 50.

6. John W. De Forest, *History of the Indians of Connecticut: From the Earliest Known Period to 1850* (1851; repr., Hamden, CT: Archon, 1964), 166–67, 177.

7. William Turnbaugh, "Assessing the Significance of European Goods in Seventeenth-Century Narragansett Society," in *Ethnohistory and Archaeology: Approaches to Postcontact Change in the Americas*, ed. J. Daniel Rogers and Samuel M. Wilson (New York: Plenum, 1993), 134–35; Constance A. Crosby, "From Myth to History, or Why King Philip's Ghost Walks Around," in *The Recovery of Meaning: Historical Archaeology in the Eastern United States*, ed. Mark P. Leone and Parker B. Potter Jr. (Washington: Smithsonian Institution Press, 1988), 183, 187.

8. McIntyre, *William Pynchon*, 12–13; Margaret M. Bruchac, "Historical Erasure and Cultural Recovery: Indigenous People in the Connecticut River Valley" (PhD diss., University of Massachusetts, 2007), 1; Peter A. Thomas, *In the Maelstrom of Change: The Indian Trade and Cultural Process in the Middle Connecticut Valley, 1635–1665* (New York: Garland, 1990), 3.

9. Thomas, *Maelstrom of Change*, 263. See also Stephen Innes, *Labor in a New Land: Economy and Society in Seventeenth-Century Springfield* (Princeton, NJ: Princeton University Press, 1983), 30. According to Thomas, Pynchon's monopoly in western New England only meant that he "controlled one percent of the Northeastern fur market" (263–64).

10. Some of Pynchon's entries from this period are illegible, so it is possible there were additional mirrors not counted. See John Pynchon, "Account Books of Major John Pynchon, Springfield, Massachusetts," vol. 1 (1651–55) and vol. 2 (1657–66) [microfilm reel 1] (Springfield: Connecticut Valley Historical Museum, 1957); and Carl Bridenbaugh and Juliette Tomlinson, eds., *The Pynchon Papers*, vol. 2, *Selections from the Account Books of John Pynchon, 1651–1697* (Boston: Publication of the Colonial Society of Massachusetts, 1985), 77, 92–93.

11. That Pynchon had so many more metal mirrors than glass suggests that white demand for the newer reflective technology may have relegated metal mirrors to trade with Native peoples. It is also possible that the larger supply of metal mirrors reveals a Native American preference for metal, but there is no additional evidence to support this explanation.

12. Glass mirrors were eclipsing metal mirrors both in language and in use. "Looking glass" or simply "glass" was becoming the generic name for reflective devices, more and more of which were actually made of glass. But not all mirrors were yet made of glass; metal mirrors were still in production and distribution. This period of transition before metal mirrors disappeared made it possible that sometimes, as in Pynchon's records, a metal mirror was called a "glass." See chapter 1 for a fuller discussion.

13. Pynchon, "Account Books of Major John Pynchon," vols. 1 and 2; Bridenbaugh and Tomlinson, *The Pynchon Papers*, 2:77, 92–93.

14. See Timothy J. Kent, *Ft. Pontchartrain at Detroit: A Guide to the Daily Lives of Fur Trade and Military Personnel, Settlers, and Missionaries at French Posts* (Ossineke, MI: Silver Fox, 2001), 2:750. Medieval circular mirrors were sometimes housed in square cases, however. See, e.g., "A Game of Chess" mirror case (ca. 1300), Victoria and Albert Museum, http://collections.vam.ac.uk/item/O88470/a-game-of-chess-mirror-case-unknown.

15. See Robert S. Grumet, *Historic Contact: Indian People and Colonists in Today's Northeastern United States in the Sixteenth through Eighteenth Centuries* (Norman: University of Oklahoma Press, 1995), 97; and Thomas, *Maelstrom of Change*, 374. Thomas does not indi-

cate whether these were made of convex or flat glass, although their circular shape and small size—as well as the early date—suggests they were convex (374).

16. Pynchon, "Account Books of Major John Pynchon," vols. 1 and 2; Fort Albany Account Books, 1706–1717 (B.3/d/16-25 [1M408]), Hudson's Bay Company Archives, a division of the Archives of Manitoba.

17. Patricia E. Rubertone, *Grave Undertakings: An Archaeology of Roger Williams and the Narragansett Indians* (Washington: Smithsonian Institution Press, 2001), 6–7, 14.

18. William Simmons, *Cautantowwit's House: An Indian Burial Ground on the Island of Conanicut in Narragansett Bay* (Providence, RI: Brown University Press, 1970), 50; Rubertone, *Grave Undertakings*, 87, 97. See also J. Patrick Cesarini, "The Ambivalent Uses of Roger Williams's *A Key into the Language of America*," *Early American Literature* 38, no. 3 (2003): 469–94.

19. Roger Williams, *A Key into the Language of America*, in *The Complete Works of Roger Williams*, vol. 1, ed. J. H. Trumbull (1866; repr., New York: Russell and Russell, 1963), 184 (unbracketed page numbers cited).

20. Ibid. Earlier in *A Key*, Williams wrote:

Boast not proud English, of thy birth & blood,
Thy brother Indian is by birth as Good.
Of one blood God made Him, and Thee & All,
As wise, as faire, as strong, as personall. (81)

Rubertone argues that Williams believed "Native peoples were spiritually lost but that their salvation was a possibility" (*Grave Undertakings*, 102).

21. Williams, *A Key*, 184.

22. Ibid., 206.

23. Ibid, 184. In 1905 the anthropologist Charles C. Willoughby summarized: "Face painting was common with both sexes, and among the men more especially when on war raids. Soot was commonly used for black, and red earth or the powdered bark of the pine tree for red. These were the more common colors. White, yellow, and blue were also used." Charles C. Willoughby, "Dress and Ornament of the New England Indians," *American Anthropologist* 7, no. 3 (1905): 499–508, 501.

24. William Simmons noted that Roger Williams's observation about Narragansett belief in a dual soul "is consistent with ideas held by numerous Algonquin tribes both north and south of Rhode Island." According to Simmons, the two souls had very different functions: *Cowwéwonck*, "concerned with dreams and visions, could probably free itself of the body during sleep, hallucinations, and daydreams. *Míchachunck*, whether considered as image or breath, may have controlled the individual's vital energy—the force that sustains life while the free soul roams about." Simmons, *Cautantowwit's House*, 54.

25. Williams, *A Key*, 154. Scholars have taken this passage to mean that the Narragansett connected míchachunck and European looking glasses. Literature scholar Renée Bergland writes that "the Narragansetts' association of their souls with looking glasses is troubling. At the least, it implies that Indian identity and spirituality were greatly affected by European trade." And archaeologist Kathleen Bragdon considers whether Native people in southern New England might have used the looking glass as a "symbolic substitute" in burials for míchachunck. Bragdon discusses a burial from the Fanning Road Graveyard on the Mashantucket Pequot Reservation in Ledyard, Connecticut. In the late nineteenth century, whites excavated a burial at this site that they described in this way: "A circular opening was dug in the earth, and the body placed in a sitting posture. A stake had been forced into the ground perpendicularly in front of it; a nail was driven into the stake, on

which was hung a looking-glass opposite the face of the dead, who was supposed to be a female." Bragdon argues that mirrors "probably . . . came to symbolize the 'see through' or clear knowledge of an individual's essence," and she takes the placement of the mirror in the burial at Fanning Road as evidence of this connection (35). This is the only recorded observance of a mirror being placed in a burial in such a manner. If the mirror was a physical substitute for the soul in Native southern New England burials, this raises the question of whether we should expect to see more mirrors in the burials. See Renée L. Bergland, *The National Uncanny: Indian Ghosts and American Subjects* (Hanover, NH: University Press of New England, 2000), 131; Kathleen J. Bragdon, *Native People of Southern New England, 1650–1775* (Norman: University of Oklahoma Press, 2009), 34–36; and D. Hamilton Hurd, *History of New London County, Connecticut* (Philadelphia: J. W. Lewis, 1882), 2:530, https://archive.org/details/historyofnewlondo2hurd.

26. Williams, *A Key*, 154. That Williams distinctly emphasizes twice the clarity of vision in his discussion of sight connected to looking glasses also provides a clue that the material object he was referring to produced what he considered to be a fairly accurate reflection of the observer.

27. In Ojibwe the word meaning "he/she sees" is *waabi;* the words for mirror are *waabamojichaagwaan* or *waabandizowin.* The Cree word for "see him" is *wapim;* the words for mirror are *wâpamonâpis* or *wâpamon*, and "s/he looks in the mirror and sees herself/himself" is *wâpamiw. The Ojibwe People's Dictionary*, http://ojibwe.lib.umn.edu/; *Online Cree Dictionary*, www.creedictionary.com/.

28. See Christopher L. Miller and George R. Hamell, "A New Perspective on Indian-White Contact: Cultural Symbols and Colonial Trade," *Journal of American History* 73, no. 2 (1986): 316. This article made an important contribution to the literature because it refuted the stereotype that Native peoples had been foolish to desire European products. The authors argue that Native peoples valued European glass (and other European goods) because of its "similarity to native substances." But Miller and Hamell's primary interest is in European glass beads rather than European mirrors. They accept, for example, Roger Williams's assertion that "the Narragansett Algonquians' word for soul had an affinity 'with a word signifying a looking glass, [or] a cleare resemblance'" (317), and Speck's interpretation of the meaning of the Naskapi word for mirror (see note 29 below).

Moreover, one of Miller and Hamell's most significant findings, using Arthur Ray's data on the Hudson's Bay Company's York Factory trading records, was that although Native peoples quickly embraced European items like glass beads, as European goods, especially between 1689 and 1763, came to be offered "almost universally . . . in contexts of war and other antisocial activities . . . Indian people were quick to put aside the once magical, now commonplace items." Miller and Hamell continue: "Arthur J. Ray has noted for the Hudson's Bay District that . . . trade in beads declined precipitously" during this period (327). But, in studying the trade at Hudson's Bay for this project, I began to doubt whether this decline was accurate. By way of an email conversation with Ann M. Carlos and Frank D. Lewis, authors of *Commerce by a Frozen Sea: Native Americans and the European Fur Trade* (Philadelphia: University of Pennsylvania Press, 2010), I concluded that Miller and Hamell were likely incorrect in this assertion. Carlos and Lewis found that in the 1720s, an average of 221 pounds of beads were traded per year at York Factory, but that by the 1760s, the number of pounds of beads traded had declined to 127 on average. But, during the 1750s and 1760s, the price of beads rose from 2 Made Beaver (MB) per pound to about 3.1 MB per pound, so it cost Native consumers nearly as many MB to purchase a much smaller quantity of beads. Moreover, in the 1760s, French traders accounted for around 40 percent of trade, especially in lighter items, like beads, that the French traders could easily transport. (See chapter 3 for a fuller discussion of the trade at Hudson's Bay.) It is critical to factor in both the significant increase in price for Native consumers and their ability to purchase this item as well from the French. There is no evidence, therefore, of a decline in interest in this good. Frank D. Lewis and Ann M. Carlos, email conversations with the author, June 2012; Arthur J. Ray, *Indians in the Fur Trade: Their Role as Trappers, Hunters, and Middlemen in the Lands Southwest of Hudson Bay, 1660–1870* (Toronto: University of Toronto Press, 1974), 79–87.

29. Reuben Gold Thwaites, ed., *Jesuit Relations and Allied Documents* (Cleveland, OH: Burrows Brothers, 1898), 33:192–93; John Bartram, *Observations on the Inhabitants, Climate, Soil, Rivers, Productions, Animals, and Other Matters Worthy of Notice* (London: J. Whiston and B. White, 1751), 33; Erminnie Smith, *The Myths of the Iroquois* (Washington: Government Printing Office, 1883), 68–69; Frank Speck, *Naskapi: The Savage Hunters of the Labrador Peninsula* (Norman: University of Oklahoma Press, 1977), 33, 164–65. In his work among the Naskapi, Speck noted something that had "escaped the notice of others," namely that the Naskapi word for soul—*atca´k*—was a component of the Naskapi word for mirrors. He noted that "the same word [atca´k] designates one's shadow," but translated the Naskapi word for mirrors *wa´pən‘atcakwoma´n* as "see-soul-metal." Speck, working in the early twentieth century, did not himself notice that the word chosen by the Naskapi to describe the physical object itself was metal, even though by 1900, mirrors would in almost all likelihood have been made of glass. That "metal" was a part of the word suggests that the Naskapi word for mirrors was rooted in either naturally occurring metal that was available in North America before the arrival of Europeans or in the Naskapi's experiences with European metal mirrors in the sixteenth or seventeenth centuries, either of which would have had less reflective capacity than later glass mirrors. So even if the Naskapi did mean "see-soul-metal" and not "see-shadow-metal," the object they were referring to was not, at least at first, a European looking glass but perhaps a naturally occurring reflective metal from North America or an early metal mirror from Europe. But *wa´pən‘atcakwoma´n* might also translate "see-shadow-metal," which casts an entirely different set of resonances on this term and the idea behind it. In other words, I would argue, we do not know enough from this evidence to definitively associate the Naskapi word for soul with their word for mirror.

30. Archaeologists have focused on burial sites because little evidence of occupation sites has been recovered. See Turnbaugh, "Assessing the Significance of European Goods," 144. The one exception to the similar variety of mirrors is that no metal mirrors appear in the burials, despite their seeming commonality aboveground. It is possible that these small pieces of metal were no longer recognizable as mirrors by the time archaeologists conducted their analyses.

31. See Elise M. Brenner, "Sociopolitical Implications of Mortuary Ritual Remains in 17th-Century Native Southern New England," in *The Recovery of Meaning: Historical Archaeology in the Eastern United States*, ed. Mark P. Leone and Parker B. Potter Jr. (Washington: Smithsonian Institution Press, 1988), 155; and Crosby, "From Myth to History," 183.

32. Simmons, *Cautantowwit's House*, 42; William Turnbaugh, *The Material Culture of RI-1000, A Mid-17th Century Narragansett Indian Burial Site in North Kingstown, Rhode Island* (Kingston: Department of Sociology and Anthropology, University of Rhode Island, 1984), 14; Turnbaugh, "Assessing the Significance of European Goods," 144; Susan G. Gibson, ed., *Burr's Hill: A 17th Century Wampanoag Burial Ground in Warren, Rhode Island* (Providence, RI: Haffenreffer Museum of Anthropology, 1980), 22; Foster H. Saville, "A Montauk Cemetery at Easthampton, Long Island," in *Indian Notes and Monographs* 2, no. 3 (New York: Museum of the American Indian, 1920), 71–74. The unusually specific 1728 date for the Pantigo burial site comes from two coins bearing that year found in a burial there.

33. Turnbaugh, *Material Culture of RI-1000*, 15.

34. These are not the only Native cemeteries at which glass mirrors have been recovered. But they are the only ones about which extensive information about the grave goods in the cemetery are available, so others are not mentioned in this discussion. I know of three other mirrors found in Native burials in southern New England: a small rectangular glass mirror (10 cm × 7 cm) with matting or a wooden frame at Long Pond, Rhode Island; a glass mirror found at Fanning Hill, Connecticut; and a broken glass mirror at the Taylor Farm site in Massachusetts. Dr. Kevin McBride (director of research at Mashantucket Pequot Museum and Research Center), in discussion with the author, June 2013; Hurd, *History of New London County*, 2:530; William B. Taylor, "The Taylor Farm Site," *Bulletin of the Massachusetts Archaeological Society* 43 (Oct. 1982): 43.

35. The ring was found in Burial 2, the grave of an infant. Simmons, *Cautantowwit's House*, 75.

36. The wooden mirror box was found in Burial 17, the final resting place of an adolescent girl. Rubertone, *Grave Undertakings*, 150–52, 192; Turnbaugh, *Material Culture of RI-1000*, 157.

37. The "pieces of mirror and leather/bark frame" were found in Burial 16. Gibson, *Burr's Hill*, 18.

38. Turnbaugh, *The Material Culture of RI-1000*, 100–101; Gibson, *Burr's Hill*, 18, 174. The circular iron box description from Burr's Hill states that the piece of glass was the top of the metal box, but it is difficult to imagine why a metal box of this era would have a breakable glass top. More likely is that the top—another piece of metal—is missing or had been removed before burial from what I infer was a mirror case.

39. Scholars have noted that a cultural logic can be discerned in southern New England grave goods. One of these patterns is that the gendered use of tools in life was recreated in grave good distribution: men were more likely to be buried with tools they used, while women were more likely to be buried with tools that women in southern New England used. See Michael S. Nassaney, "Native American Gender Politics and Material Culture in Seventeenth-Century Southeastern New England," *Journal of Social Archaeology* 4, no. 3 (2004): 343. The power to deflect may have been important in life, as well as in death. McBride has suggested that Native combatants wore kettle bale lugs for deflection during the Pequot War. Seventeenth-century New England Native peoples also acquired or made from European kettles several kinds of gorgets, pendants, and plates that would have had a reflective (or at least shimmery) surface that might have evoked the power of deflection. At West Ferry, for example, in the lavishly furnished grave Simmons referred to as the "chief's" grave, there was one "large breastplate of sheet brass, and numerous fragments of one and perhaps two more large brass plates." McBride, discussion; Simmons, *Cautantowwit's House*, 158. See also Martha Hamilton, *Silver in the Fur Trade, 1680–1820* (Chelmsford, MA: Martha Hamilton, 1995), 66–87.

40. All of Fowler's sites were in coastal Massachusetts with the exception of one in central Massachusetts at West Brookfield. The burials with these stones shared other qualities as well: they were "secondary cremation burials, in which red powdered ocher often was present." See William Fowler, "Magic Stones and Shamans," *Bulletin of the Massachusetts Archaeological Society* 36, nos. 3 and 4 (1975): 11, 16.

41. Fowler, "Magic Stones," 16. On the importance of a "ritual specialist" for burial rites see Nassaney, "Native American Gender Politics," 343–44.

42. Fowler, "Magic Stones," 17.

43. Bruce Trigger writes that there was a waxing and waning of burying the dead with "exotic materials" and that "in the centuries prior to the appearance of European goods,

burial ritualism utilizing exotic materials had reached one of its periodical low ebbs in this region. It would therefore appear that the arrival of these goods stimulated a revival of burial ceremonialism." Trigger and Washburn, "Entertaining Strangers," 376. See also Crosby, "From Myth to History," 188.

44. Simmons, *Cautantowwit's House*, 47, 69–159; Gibson, *Burr's Hill*, 17–21; Turnbaugh, *The Material Culture of RI-1000*, 59–67, 104–157; Saville, "A Montauk Cemetery," 75–86.

45. Simmons, *Cautantowwit's House*, 75.

46. Salisbury, introduction to *The Sovereignty and Goodness of God*, 7–8, 15–16, 35, 44.

47. Valerie I. J. Flint, "A Magic Universe," in *A Social History of England, 1200–1500*, ed. Rosemary Horrox and W. Mark Ormrod (Cambridge: Cambridge University Press, 2006), 346.

48. See Hugh Tait, "'The Devil's Looking-Glass': The Magical Speculum of John Dee," in *Horace Walpole: Writer, Politician, and Connoisseur*, ed. Warren Hunting Smith (New Haven, CT: Yale University Press, 1967), 197. John Dee's mirror was seen by the author at the British Museum in 2006. Deborah E. Harkness argues, however, that the mirror Dee used was not the one that is now housed in the British museum. See Deborah E. Harkness, *John Dee's Conversations with Angels: Cabala, Alchemy, and the End of Nature* (Cambridge: Cambridge University Press, 1999), 29–30. In either case, Dee used showstones in his practices. The obsidian mirror in the collection of the British Museum allows a modern-day observer to catch a glimpse of the kind of reflection that obsidian—a naturally occurring reflective material—offered. On the use of obsidian as a mirror in Central America see Colin McEwan and Leonardo López Luján, *Moctezuma: Aztec Ruler* (London: British Museum Press, 2009), 91, 167, 238. For a discussion of mirrors in "pre-Columbian" South and Central America more broadly see Benjamin Goldberg, *The Mirror and Man* (Charlottesville: University Press of Virginia, 1985), 79–93.

49. As the historian Stuart Clark puts it, "All over Europe, men and women practised divination with scissors and sieves, or books and keys, or by peering into the flat surfaces made by water or mirrors or 'crystals.'" Stuart Clark, "Popular Magic," in *Witchcraft and Magic in Europe: The Period of the Witch Trials*, ed. Bengt Ankarloo and Stuart Clark (Philadelphia: University of Pennsylvania Press, 2002), 103. For other mentions of reflections used in magical practices, see Wallace Notestein, *A History of Witchcraft in England from 1558–1718* (New York: Russell and Russell, 1965), 23, 211, 213; and Keith Thomas, *Religion and the Decline of Magic* (New York: Charles Scribner's Sons, 1971), 117, 186, 549.

50. Richard Bernard, *A Guide to Grand-Jury Men . . . in cases of Witchcraft* (London: Felix Kingston for Ed. Blackmore, 1627), 136–38, http://quod.lib.umich.edu/e/eebo/A09118.0001.001?view=toc.

51. Increase Mather, *A Further Account of the Tryals of the New-England Witches* (London: J. Dunton, 1693), 283, www.gutenberg.org/files/28513/28513.txt; Cotton Mather, *Memorable Providences, Relating to Witchcrafts and Possessions*... (London: Tho. Parkhurst, 1691), 117; Increase Mather, *Angelographia* (Boston: B. Green and J. Allen, 1696), 25–26. Not everything mentioned in this context as a "divining glass" should be assumed to have been a mirror or looking glass, although some undoubtedly were. There was another kind of divining glass known as a "Venus glass," which was a glass of water in which an egg white was suspended and its shape in the water "read," often to tell the future. Seventeenth-century Puritan minister John Hale told of one such incident: "I knew one of the Afflicted persons, who (as I was credibly informed) did try with an egg and a glass to find her future Husbands Calling; till there came up a Coffin, that is, a Spectre in likeness of a Coffin." Two venus glasses were also broken during a church meeting at Increase Mather's church in Boston in 1687 and a venus glass was mentioned in the examination of Sarah Cole during the Salem

Witchcraft Trials. See John Hale, *A Modest Enquiry into the Nature of Witchcraft* (Boston: B. Green and J. Allen, 1702), 132–33, http://name.umdl.umich.edu/N00872.0001.001; Samuel Sewall, *Diary*, in *Collections of the Massachusetts Historical Society* (Boston: Massachusetts Historical Society, 1878), 5:183, https://archive.org/details/diaryofsamuelsew01sewaiala; and "Examination of Sarah Cole," Salem Witch Trials: Documentary Archive and Transcription Project, http://salem.lib.virginia.edu/texts/tei/BoySalCombined?div_id=n34.

52. Salisbury, introduction to *The Sovereignty and Goodness of God*, 38–40.

53. Sabine Melchior-Bonnet, *The Mirror: A History*, trans. Katharine H. Jewett (New York: Routledge, 2001), 200.

54. Cotton Mather, *Ornaments for the Daughters of Zion; or, The Character and Happiness of a Vertuous Woman* (Cambridge: S. G. and B. G. for Samuel Phillips at Boston, 1692), 8, http://quod.lib.umich.edu/e/evans/N00500.0001.001?view=toc; *Bethiah: The Glory Which Adorns the Daughters of God. And the Piety, Wherewith Zion Wishes to See Her Daughters Glorious* (Boston: J. Franklin, 1722), 50; Kenneth Silverman, *The Life and Times of Cotton Mather* (New York: Harper and Row, 1984), 395.

55. Joseph Rowlandson of Wethersfield Probate Inventory, 1678, case no. 4658 (Hartford, Connecticut Probate District), Connecticut State Library Microfilm.

56. Kevin M. Sweeney, "Furniture and the Domestic Environment in Wethersfield, Connecticut, 1639–1800," in *Material Life in America, 1600–1860*, ed. Robert Blair St. George (Boston: Northeastern University Press, 1988), 264.

57. James Axtell, "The First Consumer Revolution," in *Consumer Society in American History: A Reader*, ed. Lawrence B. Glickman (Ithaca, NY: Cornell University Press, 1999), 87.

CHAPTER 3: Looking-Glass Ownership in Early America

1. The *Agenoria* began its voyage into Africa at Cape Formoso on the River Niger (present-day Nigeria). This cape is situated between the Bight of Biafra and the Bight of Benin, two regions that were the points of embarkation for 28.7 percent of the estimated 12,521,000 captive Africans sold into slavery between 1501 and 1867. See John R. Bartlett, *Letter of Instructions to the Captain and the Supercargo of the Brig "Agenoria," Engaged in a Trading Voyage to Africa* ([S.l.]: privately printed for Howard Greene and Arnold G. Talbot, 1933), 4; and David Eltis and David Richardson, *Atlas of the Transatlantic Slave Trade* (New Haven, CT: Yale University Press, 2010), 15 (map 9).

2. Bartlett, *Letter of Instructions*, 16, 20, 31, 32, 37, 40. Members of the crew purchased a "Small Looking Glass," valued at $0.50; a "large looking glass," valued at $0.67; another "looking glass," also valued at $0.67; and a "looking glass," valued at $1.67. It is not known whether these glasses were taken from the trading stock or if a separate set of supplies was available for crew purchases. The trading records listed only two kinds of looking glasses, but the crew purchased mirrors at three different prices. This suggests at least that the crew had more choices than the African consumers did.

3. See Holly V. Izard, "Random or Systematic? An Evaluation of the Probate Process," *Winterthur Portfolio* 32, no. 2/3 (1997): 147–67, 153–54.

4. Eltis and Richardson, *Atlas of the Transatlantic Slave Trade*, 15. Of the estimated 12,521,000 Africans who were captured and forced into slavery, 6 percent departed from Senegambia, 3.1 percent from Sierra Leone, 2.7 percent from Windward Coast, 9.7 percent from Gold Coast, 16 percent from Bight of Benin, 12.7 percent from Bight of Biafra, and 45.5 percent from West Central Africa. The remaining 4.3 percent departed from Southeast Africa, a geographic region beyond the scope of this project.

5. The nations discussed here were all located on the coast of West Africa, which encompassed the embarkation areas of Senegambia, Sierra Leone, Windward Coast, Gold Coast, Bight of Benin and Bight of Biafra. Of those captive Africans taken to North America, 23.7 percent came from Senegambia, 11.4 percent from Sierra Leone, 5.5 percent from Windward Coast, 14.2 percent from Gold Coast, 2.3 percent from Bight of Benin, and 17.6 percent from Bight of Biafra. See Eltis and Richardson, *Atlas of the Transatlantic Slave Trade*, 15 (map 9), 205 (map 137).

6. Pieter de Marees, *Description and Historical Account of the Gold Kingdom of Guinea*, trans. and ed. Albert. van Dantzig and Adam Jones (1602; repr., New York: Oxford University Press, 1987), 53, 63, 83; Andreas Joshua Ulsheimer, "Voyage of 1603–4," in *German Sources for West African History, 1599–1669*, ed. Adam Jones (Wiesbaden: Franz Steiner, 1983), 29, 34; Wilhelm Johann Müller, "Description of the Fetu Country, 1662–69," in Jones, *German Sources for West African History, 1599–1669*, 247.

7. See Eleanor S. Godfrey, *The Development of English Glassmaking, 1560–1640* (Chapel Hill: University of North Carolina Press, 1975), 238–39. The 1640 shipment was also destined for a third place, noted only as "?Robordence."

8. Jean Barbot, *Barbot on Guinea: The Writings of Jean Barbot on West Africa, 1678–1712*, vol. 1, ed. P. E. H. Hair, Adam Jones, and Robin Law (London: Hakluyt Society, 1992), 105; *Olfert Dapper's Description of Benin*, ed. Adam Jones (1668; repr., Madison: African Studies Program, University of Wisconsin-Madison, 1998), 19.

9. John Atkins, *A Voyage to Guinea, Brasil, and the West-Indies* (London: Ward and Chandler, 1735), 162, 166.

10. John Matthews, *A Voyage to the River Sierra Leone, on the Coast of Africa* (London: B. White and Son, 1791), 115.

11. Eltis and Richardson, *Atlas of the Transatlantic Slave Trade*, map 9 (15), map 11 (18–19), maps 92–97 (136–39), map 137 (205), and 307. Less is known about the exact points of origin of captive Africans brought from the interior of the continent to the coast of West Central Africa.

12. See Filipa Ribeiro da Silva, *Dutch and Portuguese in Western Africa: Empires, Merchants and the Atlantic System, 1580–1674* (Leiden, The Netherlands: Brill, 2011), 207.

13. Dixon Denham, Hugh Clapperton, and Walter Oudney, *Narrative of Travels and Discoveries in Northern and Central Africa, in the Years 1822, 1823, and 1824*, 2 vols. (London: John Murray, 1828), 2:189–90, 1:301; "Map of Denham and Clapperton's Journey," in Jules Verne, *Celebrated Travels and Travellers, Part III: The Great Explorers of the Nineteenth Century*, trans. N. D'Anvers (London: Sampson Low, Marston, Searle, and Rivington, 1881), www.gutenberg.org/files/26658/26658-h/26658-h.htm#ill26.

14. Denham, Clapperton, and Oudney, *Narrative of Travels and Discoveries*, 2:328.

15. Despite the people's interest in mirrors, Denham did not distribute any looking glasses to the residents of Yeddie, likely because he did not have any, something he revealed a few days later when he encountered that sultan who wanted one. Ibid., 1:269–70.

16. Ibid., 2:299.

17. Hugh Clapperton and Richard Lander, *Journal of a Second Expedition into the Interior of Africa from the Bight of Benin to Soccatoo* (Philadelphia: Carey, Lea and Carey, 1829), 81, 121, 242, 261.

18. Richard and John Lander, *Journal of an Expedition to Explore the Course and Termination of the Niger*, 2 vols. (New York: Harper and Brothers, 1839, 1844), 1:149, 203, 238, 266–67 (https://archive.org/details/journalofexpedit01land), and 2:63, 73, 79, 213–14, 330–1 (https://archive.org/details/journalofexpedit02rich). They also discovered upon arrival that

many of the needles had a manufacturing error (they were "eyeless") rendering them useless. They ran so low on gifts that they resorted to parting with the buttons from their own garments (1:266–7). For more on shards of mirror glass adorning textiles, see Graham Child, "Mirror," in *The Dictionary of Art*, vol. 21, *Medallion to Montalbani* (New York: Grove, 1996), 717.

19. Igbo captives in the African slave trade are included in the Bight of Biafra region, from which arrived 17.6 percent of the enslaved Africans coming to North America. Herbert M. Cole and Chike C. Aniakor, *Igbo Arts: Community and Culture* (Los Angeles: Museum of Cultural History, 1984), 1; Michael A. Gomez, *Exchanging our Country Marks: The Transformation of African Identities in the Colonial and Antebellum South* (Chapel Hill: University of North Carolina Press, 1998), 114; Eltis and Richardson, *Atlas of the Transatlantic Slave Trade*, 205 (map 137), 212 (map 141).

20. MacGregor Laird and R.A.K. Oldfield, *Narrative of an Expedition into the Interior of Africa*, 2 vols. (London: Richard Bentley, 1837), 1:384–85. For a different interpretation, see Ann Smart Martin, *Buying into the World of Goods: Early Consumers in Backcountry Virginia* (Baltimore: Johns Hopkins University Press, 2008), 189.

21. Laird and Oldfield, *Narrative*, 2:181; Cole and Aniakor, *Igbo Arts*, 54–55.

22. The nineteenth- and early twentieth-century collection practices evidenced in the Pitt Rivers collection make it impossible to date the artifacts in the collection. See William Ryan Chapman, "Arranging Ethnology: A. H. L. F. Pitt Rivers and the Typological Tradition," in *Objects and Others: Essays on Museums and Material Culture*, ed. George W. Stocking Jr. (Madison: University of Wisconsin Press, 1985), 15–48.

23. The two mirrors of this type that had been collected in 1925 (1925.12.3 and 1925.53.7.1–3) were both noted as having been used for divination. Other objects with glass mirrors in the Pitt Rivers collections from West Africa include two small figurines (1884.65.65.1 and 1884.65.65.2), noted as "possibly *nkisi* figures with remains of 'medicine'" (see chapter 6 for a discussion of nkisi); a mirror incorporated into a religious object made primarily of wicker, possibly from Nigeria (1884.56.48); and a mask from southern Nigeria (1914.26.8).

24. This archaeological archive includes fifty-five excavations that took place at twenty-one plantation sites. Mirror glass was recovered from twenty-three excavations at eleven of these sites: Monticello, Palace Lands, Richmond, Poplar Forest, Utopia, Ashcombs, Fairfield, Stagville, Middleburg, Yaughan, and the Hermitage. Artifact Query 3, Digital Archaeological Archive of Comparative Slavery, Thomas Jefferson Foundation, www.daacs.org/.

25. Kathleen A. Parker and Jacqueline L. Hernigle, *Portici: Portrait of a Middling Plantation in Piedmont*, Virginia Occasional Report no. 3, Regional Archeology Program, National Capital Region (Washington: National Park Service, 1990), 132, 137; K. L. Brown, "Material Culture and Community Structure: The Slave and Tenant Community at Levi Jordan's Plantation, 1848–1892," in *Working Toward Freedom: Slave Society and Domestic Economy in the American South*, ed. L. E. Hudson Jr. (New York: University of Rochester Press, 1994): 95–118; Christopher C. Fennell, "BaKongo Identity and Symbolic Expression in the Americas," in *Archaeology of Atlantic Africa and the African Diaspora*, ed. Akinwumi Ogundiran and Toyin Falola (Bloomington: Indiana University Press, 2007), 221.

26. John Michael Vlach, *By the Work of Their Hands: Studies in Afro-American Folklife* (Ann Arbor: UMI Research Press, 1991), 61.

27. Guion Griffis Johnson, *Ante-Bellum North Carolina: A Social History* (Chapel Hill: University of North Carolina Press, 1937), 496, http://docsouth.unc.edu/nc/johnson/johnson.html; "George Fleming," in *The American Slave: A Composite Autobiography*, ed.

George P. Rawick, supp. ser. 1, vol. 11, *North Carolina and South Carolina Narratives* (Westport, CT: Greenwood, 1977), 129–30; Eve Scurlock, Testator, Abstracts of Wills on File in the Surrogate's Office, vol. 5 (1754–60), *Collections of the New-York Historical Society* (New York, 1897), 42.

28. Although there were many beliefs about the bad luck associated with breaking a looking glass, people did not seem to reject using looking glasses that had been previously broken. Perhaps the bad luck only attached to the person who broke the glass, not to those who later used the broken pieces.

29. Charles L. Perdue Jr., Thomas E. Barden, and Robert K. Phillips, eds., *Weevils in the Wheat: Interviews with Virginia Ex-slaves* (Charlottesville: University Press of Virginia, 1976), 227; Frances Anne Kemble, *Journal of a Residence on a Georgia Plantation in 1838–1839* (New York: Harper and Brothers, 1863), 261; Jacob D. Green, *Narrative of the Life of J. D. Green, a Runaway Slave, from Kentucky . . .* (Huddersfield, UK: Henry Fielding, Pack Horse Yard, 1864), 11–12, http://docsouth.unc.edu/neh/greenjd/greenjd.html; Anne Maddox interview, *Slave Narratives: A Folk History of Slavery in the United States from Interviews with Former Slaves* (hereafter *SN*), vol. 1, *Alabama Narratives* (Washington: Federal Writers' Project, 1941), 273. The *Slave Narratives* collection is available at the Library of Congress website: www.loc.gov/collections/slave-narratives-from-the-federal-writers-project-1936-to-1938/about-this-collection/.

30. There is an extensive, and growing, literature on the slaves' internal economy. See Philip Morgan, "Work and Culture: The Task System and the World of Lowcountry Blacks, 1700–1800," *William and Mary Quarterly* 39, no. 4 (1982): 563–99; Philip Morgan, "The Ownership of Property by Slaves in the Mid-Nineteenth-Century Low Country," *Journal of Southern History* 49, no. 3 (1983): 399–420; Roderick McDonald, *The Economy and Material Culture of Slaves: Goods and Chattels on the Sugar Plantations of Jamaica and Louisiana* (Baton Rouge: Louisiana State University Press, 1993); Barbara Heath, *Hidden Lives: The Archaeology of Slave Life at Thomas Jefferson's Poplar Forest* (Charlottesville: University Press of Virginia, 1999), 50–58; Dylan C. Penningroth, *The Claims of Kinfolk: African American Property and Community in the Nineteenth-Century South* (Chapel Hill: University of North Carolina Press, 2003); and Kathleen M. Hilliard, *Masters, Slaves, and Exchange: Power's Purchase in the Old South* (New York: Cambridge University Press, 2013).

31. Martin, *Buying into the World of Goods*, 180; Charles B. Dew, *Bond of Iron: Master and Slave at Buffalo Forge* (New York: Norton, 1995), 180–81; J. Miller M'Kim, *The Freedmen of South Carolina: An Address Delivered by J. Miller M'Kim, in Sansom Hall, July 9th, 1862. Together with a Letter from the Same to Stephen Colwell, Esq., Chairman of the Port Royal Relief Committee* (Philadelphia: Willis P. Hazard, 1862), 21.

32. "Confession of John Joyce, Alias Davis, Who Was Executed on Monday, the 14th of March, 1808. For the Murder of Mrs. Sarah Cross; With an Address to the Public and People of Colour" (Philadelphia: Printed for the Benefit of Bethel Church, 1808), 15, http://docsouth.unc.edu/neh/joyce/joyce.html; Hilliard, *Masters, Slaves, and Exchange*, 107–9.

33. Pauline Johnson and Felice Boudreaux interview, *SN*, vol. 16, *Texas Narratives, Part 2*, 226; Daphne Williams interview, *SN*, vol. 16, *Texas Narratives, Part 4*, 160; Charlotte Beverly interview, *SN*, vol. 16, *Texas Narratives, Part 1*, 84; Jeff Calhoun interview, ibid., 189; Harriet Collins interview, ibid., 242; Van Moore interview, *SN*, vol. 16, *Texas Narratives, Part 3*, 129. For a fuller discussion of the Works Progress Administration interviews, see chapter 5.

34. "Millie Manuel," in *The American Slave: A Composite Autobiography*, ed. George P. Rawick, supp. ser. 2, vol. 7, pt. 6, *Texas Narratives* (Westport, CT: Greenwood, 1979), 25–70; Alexander Robertson interview, *SN*, vol. 14, *South Carolina Narratives, Part 4*, 32.

35. *The Spirit Sings: Artistic Traditions of Canada's First Peoples* (Toronto: McClelland and Steward; Calgary: Glenbow Museum, 1987), map on 8–9.

36. Ann M. Carlos and Frank D. Lewis, *Commerce by a Frozen Sea: Native Americans and the European Fur Trade* (Philadelphia: University of Pennsylvania Press, 2010), 2–4.

37. See Arthur J. Ray, *Indians in the Fur Trade: Their Role as Trappers, Hunters, and Middlemen in the Lands Southwest of Hudson Bay, 1660–1870* (Toronto: University of Toronto Press, 1974), 61–62.

38. E. E. Rich, *Minutes of the Hudson's Bay Company, 1671–1674* (Toronto: Champlain Society, 1942), 100, 108; E. E. Rich, *Minutes of the Hudson's Bay Company, 1679–1684*, pt. 1, *1679–1682* (Toronto: Champlain Society, 1945), 212.

39. One of these men was the Frenchman Peter Radisson. He had first arrived in North America in 1651 and had been active in the fur trade for many years by this time. See Peter C. Newman, *Empire of the Bay: The Company of Adventurers That Seized a Continent* (New York: Penguin, 2000), 54–56; and Peter Radisson, *Voyages of Peter Esprit Radisson* (London: Publications of the Prince Society, 1858), www.gutenberg.org/ebooks/6913.

40. Ray, *Indians in the Fur Trade*, 67. The mirrors this trader brought to display to potential customers were leather looking glasses.

41. The only other scholar I have found who even noticed this item in the Hudson's Bay Company records was Toby Morantz, who, not surprisingly, given their very unusual name, mistakenly identifies these mirrors as "tin skows" without further commentary. See Toby Morantz, "The Fur Trade and the Cree of James Bay," in *Old Trails and New Directions: Papers of the Third North American Fur Trade Conference*, ed. Carol M. Judd and Arthur J. Ray (Toronto: University of Toronto Press, 1980), 42.

42. Fort Albany Account Books, 1706–17 (B.3/d/16-25 [1M408]), Hudson's Bay Company Archives, a division of the Archives of Manitoba. For more information on the archives see www.gov.mb.ca/chc/archives/hbca/index.html.

43. Ray, *Indians in the Fur Trade*, 51.

44. Ibid.; Carlos and Lewis, *Commerce by a Frozen Sea*, 42, 72. The data on mirrors for Fort Albany are from Fort Albany Account Books, 1692–96 (B.3/d/1-7 [1M406]), 1697–1706 (B.3/d/8-15 [1M407]), and 1706–17 (B.3/d/16-25 [1M408]), Hudson's Bay Company Archives. The data from York Factory, 1716–70, are from Carlos and Lewis, *Commerce by a Frozen Sea*, 81–85. Economics professors Carlos and Lewis kindly supplied me with annual data that did not appear in their published volume. The data from York Factory, 1775–82, are from York Factory Accounts Books, 1774–82 (B.239/d/65-72 [1M672], Hudson's Bay Company Archives. The population estimate for York Factory's hinterland is from Carlos and Lewis, *Commerce by a Frozen Sea*, 72. I developed the population estimate for Albany's hinterland using several data points. Carlos and Lewis note that Fort Albany shared a 230,000-square-mile hinterland with the much smaller Fort Moose during the first half of the eighteenth century (41–43). Carlos and Lewis's population estimate of one person per fifty square miles suggests that there were forty-six hundred people trading at Fort Albany and Fort Moose (72). Together, Fort Albany and Fort Moose traded 34,500 MB annually. Fort Albany traded an average of 26,000 MB (or 75 percent of the total) annually, while Fort Moose traded 8,500 MB (or 25 percent of the total), so we can estimate that Fort Albany had 75 percent of the population, or 3,450 people. The number of MB traded annually at Fort Moose and Fort Albany is from Arthur Ray and Donald Freeman, *"Give Us Good Measure": An Economic Analysis of Relations between the Indians and the Hudson's Bay Company before 1763* (Toronto: University of Toronto Press, 1978), 50.

45. A third potentially mitigating factor does not appear to have played a role in the distribution of mirrors. Many goods entered Native hands in these regions as gifts from the

Hudson's Bay Company, which was part of the process of trade. These gifts could be quite substantial. Toby Morantz notes that gifts "sometimes represented as much as half the value of the furs brought, though more usually about one-fifth." I have found no evidence, however, that mirrors were included in these gift exchanges. See Morantz, "The Fur Trade and the Cree of James Bay," 53.

46. Ray, *Indians in the Fur Trade*, 87–88.

47. E. E. Rich, *The History of the Hudson's Bay Company, 1670–1870*, vol. 1, *1670–1763* (London: Hudson's Bay Record Society, 1958), 116–32; Carlos and Lewis, *Commerce by a Frozen Sea*, 42, 58–62; Ray, *Indians in the Fur Trade*, 52–53; Frank Lewis, email message to author, June 14, 2012; for the standard of trade at Fort Albany in 1733, see www.hbcheritage.ca/hbcheritage/history/business/fur/standardtrade1733.asp.

48. Ray, *Indians in the Fur Trade*, 79–87.

49. York Factory Accounts Books 1774–82 (B.239/d/65-72 [1M672]), Hudson's Bay Company Archives.

50. Jeffrey P. Brain, *Tunica Treasure* (Cambridge, MA: Peabody Museum of Archaeology and Ethnology, Harvard University, 1979), 275–79, 291; Jeffrey P. Brain, *Tunica Archaeology* (Cambridge, MA: Peabody Museum Press, 1988), 405; Inventory of Samuel Eveleigh's Personal Possessions and Store Goods, Records of the Secretary of State, Recorded Instruments, Inventories of Estates, 1736, South Carolina Department of Archives and History, Columbia; Dean Anderson, "The Flow of European Trade Goods into the Western Great Lakes Region, 1715–1760," in *The Fur Trade Revisited: Selected Papers of the Sixth North American Fur Trade Conference, Mackinac Island, Michigan, 1991*, ed. Jennifer S. H. Brown, W. J. Eccles, and Donald P. Heldman (East Lansing: Michigan State University Press, 1994), 93–115.

51. Timothy J. Kent, *Ft. Pontchartrain at Detroit: A Guide to the Daily Lives of Fur Trade and Military Personnel, Settlers, and Missionaries at French Posts* (Ossineke, MI: Silver Fox, 2001), 2:749.

52. "Report of Indian Councils at Detroit," in *Report of the Pioneer Society of the State of Michigan*, vol. 9 (Lansing, MI: Wynkoop Hallenbeck Crawford, State Printers, 1908), 471.

53. York Factory Accounts Books, 1774–82 (B.239/d/65-72 [1M672]), Hudson's Bay Company Archives; Eric Jay Dolin, *Fur, Fortune, and Empire: The Epic History of the Fur Trade in America* (New York: Norton, 2010), 194, 221, 266; American Fur Company Papers, roll 11 (M151), Minnesota Historical Society, St. Paul.

54. "Tabeau, Pierre-Antoine," in Elin Woodger and Brandon Toropov, *Encyclopedia of the Lewis and Clark Expedition* (New York: Facts on File, 2004), 335; Pierre-Antoine Tabeau, *Tabeau's Narrative of Loisel's Expedition to the Upper Missouri*, ed. Annie Heloise Abel (Norman: University of Oklahoma Press, 1939), 171; Margaret A. Frink, "Adventures of a Party of Gold-Seekers," in *Best of Covered Wagon Women*, ed. Kenneth Holmes (Norman: University of Oklahoma Press, 2008), 69–70.

55. Prince Alexander Philip Maximilian zu Wied-Neuwied, *Early Western Travels, 1748–1846*, vol. 23, pt. 2, *Travels in the Interior of North America*, ed. Reuben Gold Thwaites (Cleveland, OH: A. H. Clark, 1906), 258. Maximilian and Swiss painter Karl Bodmer had traveled to North America to study Native American life. See Robert J. Moore, *American Indians: The Art and Travel of Charles Bird King, George Catlin, and Karl Bodmer* (Vermicelli, Italy: VMB, 2008), 202, 268.

56. For more information about the probate inventories included in this study, please contact the author. Although probate inventories provide a wealth of information for scholars interested in material culture, they do present challenges. In addition to those already noted, historian Gloria Main's analysis of how probate records could be used as a historical

source cautioned that they provide a distorted view of the population because "most probated decedents were males who tended to be older and richer than their neighbors still living." Holly Izard's study of both probated and unprobated decedents in Sturbridge, Massachusetts, attenuated Main's concern about the distortion caused by wealth by exploring why some very poor people appeared in the probate system. Izard found that there were five criteria that started the probate process. Estates entered probate if the decedent had written a will, owned real property, left minor heirs, was owed money, or died in debt. Not everyone who met one of these requirements entered probate, but people who did not meet one of these criteria did not. These factors, not specifically age or wealth, determined entry into the probate process. Izard concluded that probate records do allow us to "gauge availability of goods in a particular time and place . . . and how the local material landscape changed over time," which is the use that will be made of the probate inventories analyzed for this study. See Gloria Main, "Probate Records as a Source for Early American History," *William and Mary Quarterly* 32, no. 1 (1975): 89–99, 96; and Izard, "Random or Systematic?" 167.

57. See Alice Hanson Jones, *American Colonial Wealth: Documents and Methods*, 3 vols. (New York: Arno, 1977). The complete Jones collection contains 919 inventories filed in 1774 from Massachusetts, Connecticut, Pennsylvania, New York, New Jersey, Delaware, Maryland, Virginia, North Carolina, and South Carolina, of which 836 had usable data for the purposes of this study. I did not use any of the New Jersey probate inventories (N=25) because household goods were often lumped together. In determining mirror ownership for each quartile in 1774 using the Jones data, I also had to exclude the North Carolina inventories because they often did not list the value of individual goods.

58. See Richard Middleton, *Colonial America: A History, 1585–1776*, 2nd ed. (Oxford: Blackwell, 1996), 406.

59. By 1860, Suffolk County had a population of 192,700, New York County had a population of 813,669, and Charleston County had a population of 70,100. See Joseph C. G. Kennedy, *Population of the United States in 1860; Compiled from the Original Returns of the Eighth Census, under the Direction of the Secretary of the Interior* (Washington: Government Printing Office, 1864), www.census.gov/prod/www/decennial.html.

60. Suffolk County, Massachusetts, Wills and Inventories of Estates, 1628–1852 [1660–62, 1695–97], Winterthur Library, Joseph Downs Collection of Manuscripts and Printed Ephemera, Winterthur, Delaware. Charleston, South Carolina Inventories of Estates, vol. 52: 1687–1710; vol. 53: 1692–93; vol. 54: 1694–1704; vol. 55: 1711–18; vol. 56: 1714–17; vol. 57: 1716–21, South Carolina Department of Archives and History.

61. This disparity between rural and urban rates of ownership of looking glasses is similar to Laurel Thatcher Ulrich's finding that in 1700 in rural Essex, Massachusetts, looking-glass ownership was 31 percent, while among twelve commercially oriented Salem, Massachusetts, residents (more comparable to Suffolk County, MA, and Charleston County, SC, in this study), it had already reached 75 percent. See Laurel Thatcher Ulrich, *Good Wives: Image and Reality in the Lives of Women in Northern New England, 1650–1750* (New York: Vintage, 1991), 17.

62. Hampshire County, Massachusetts, Probate Records, 1660–1820 [1678–97, 1709–15], Winterthur Library. Kevin Sweeney's study of Wethersfield, Connecticut, found a similarly significant increase in mirror ownership at the beginning of the eighteenth century. The town of Wethersfield is also located on the Connecticut River, thirty miles south of Springfield, MA (Hampshire County). Between 1636 and 1670 only two Wethersfield residents owned a looking glass—John Lattamore (1662), who had the most valuable estate that entered probate during this period, and Thomas Lord (1662), "the town's schoolmaster and

the offspring of a wealthy Hartford mercantile family" (266). In the period 1671 to 1720, "over two-thirds of the looking glasses were found in estates ranking among the wealthiest one-fourth of the estates probated during this period" (268). Overall rates of mirror ownership in Wethersfield increased from 12 percent between 1681 and 1690 to 26 percent during the period 1691 to 1700. Averaged together, rates of ownership during these twenty years in Wethersfield were at 19 percent compared to 13 percent between 1678 and 1697 in Hampshire County. The lower percentage in Hampshire may be attributed to the additional three years Sweeney included in his study (1698–1700; these extra years may have been key because mirror ownership was rising quickly at the beginning of the eighteenth century in Hampshire County), as well as the difference between mirror ownership in one town (Wethersfield) vs. an entire county (Hampshire, which would have been more rural overall). By the decade of the 1710s in Wethersfield, however, overall mirror ownership reached 38 percent, compared to 40 percent in Hampshire County. See Kevin M. Sweeney, "Furniture and the Domestic Environment in Wethersfield, Connecticut, 1639–1800," in *Material Life in America, 1600–1860*, ed. Robert Blair St. George (Boston: Northeastern University Press, 1988), 261–68.

63. Hamilton (1712–56) had immigrated to Maryland in 1738. Alexander Hamilton, *Gentleman's Progress: The Itinerarium of Dr. Alexander Hamilton, 1744*, ed. Carl Bridenbaugh (Chapel Hill: University of North Carolina Press, 1948), 55.

64. T. H. Breen, "The Meaning of Things: Interpreting the Consumer Economy in the Eighteenth Century," in *Consumption and the World of Goods*, ed. John Brewer and Roy Porter (New York: Routledge, 1997), 249–60, 251–52; *South-Carolina Gazette*, Jan. 17, 1736, quoted in Breen, "The Meaning of Things," 258.

65. Josiah Quincy Jr., *Memoir of the Life of Josiah Quincy, Junior of Massachusetts: 1744–1775*, ed. Susan Quincy (Boston: John Wilson and Son, 1874), 72–73; Josiah Quincy Jr., "Journal of Josiah Quincy, Junior, 1773," ed. Mark A. De Wolfe Howe, *Proceedings of the Massachusetts Historical Society* 49 (June 1916): 444–45; Historic Charleston Foundation, *Grandeur Preserved: The House Museums of Historic Charleston Foundation* (Charleston: History Press, 2008), 13.

66. Probate Inventories of Josiah Baker and James St. John, 1743, Charleston County Inventories, vol. 71, 246, 263, South Carolina Department of Archives and History. Nancy Carlisle found the same pattern of mirror placement in her study of one hundred eighteenth-century probate inventories from Boston and Philadelphia. See Nancy Camilla Carlisle, "A Reflection of the Times: The Looking Glass in Eighteenth-Century America" (master's thesis, University of Delaware, 1983), 47–50.

67. Suffolk County, Wills and Inventories of Estates, 1628–1852 [1813], Winterthur Library.

68. Charleston, South Carolina, Inventories of Estates, vol. 71, 1739–43, South Carolina Department of Archives and History; Suffolk County, Wills and Inventories of Estates, 1628–1852 [1695–97], Winterthur Library; Hampshire County, Probate Records, 1660–1820 [1741–45], Winterthur Library; New York City and Vicinity, Inventories of Estates, 1717–1844 [1761–90], New-York Historical Society. Settlement patterns in Hampshire County, Massachusetts, continued to be rural compared to Suffolk County during the eighteenth century. In the 1765 census Suffolk County's Boston had 15,520 inhabitants, and of the county's eighteen towns, nine had more than one thousand inhabitants. In contrast, the largest town in Hampshire County, Springfield, had 2,755 inhabitants, and of the county's twenty-eight towns, only three had more than one thousand. See Everts B. Greene and Virginia D. Harrington, *American Population before the Federal Census of 1790* (New York: Columbia University Press, 1932), 12–46.

69. In Charleston, South Carolina, alone there were at least fifteen such craftsmen in the eighteenth century according to the Index of Early Southern Artists and Artisans, Museum of Early Southern Decorative Arts, Winston-Salem, North Carolina (hereafter IESAA).

70. Edward Weyman advertisement, *South-Carolina Gazette* (Charleston, SC), Nov. 5–12, 1763, 1.

71. George Stattler, *Charleston (SC) City Gazette, Daily Advertiser*, Dec. 4, 1797, 3-2, IESAA; John Parkinson, *South Carolina Weekly Gazette* (Charleston), March 22, 1783, 2–3, IESAA.

72. John Elliott advertisement, *Pennsylvania Gazette*, Nov. 24, 1763, quoted in Martin, *Buying into the World of Goods*, 237–38n42.

73. Edward Weyman advertisement, *South Carolina Gazette and Country Journal* (Charleston), Feb. 4, 1766, 3-2, IESAA.

74. John Elliott advertisement, *Pennsylvania Journal and Weekly Advertiser*, July 2, 1765, quoted in Carlisle, "A Reflection of the Times," 23.

75. South Carolina Department of Archives and History, Columbia, South Carolina, Records of the Secretary of State, Recorded Instruments, Inventories of Estates (WPA Transcripts), vol. 71 (1739–43).

76. Data compiled from Jones, *American Colonial Wealth*.

77. Suffolk County, Wills and Inventories of Estates, 1628–1852 [1813], Winterthur Library; Plymouth County Probate Court Records, 1686–1850 [1850], Winterthur Library.

78. Charleston County Probate Court, Book F: 1819–1824; Hampshire County, Probate Records, 1660–1820 [1813–16], Winterthur Library. New York City is not included here because no nineteenth-century probate records were located.

79. Vernon Foster, *Spartanburg: Facts, Reminiscences, Folklore* (Spartanburg, SC: Spartanburg County Foundation, 1998), 108–10; Walter Edgar, *South Carolina: A History* (Columbia: University of South Carolina Press, 1998), 286; Lacy Ford, *Origins of Southern Radicalism: The South Carolina Upcountry, 1800–1860* (New York: Oxford University Press, 1988), 46–47.

80. Record of the Office of the Probate Judge, Spartanburg County: Inventories, 1787–1810; Inventory and Appraisement Book, 1811–18 (C729); Inventory and Appraisement Book, 1818–25 (C730); Inventory and Appraisement Book, 1830–35 (C730); Inventory and Appraisement Book, 1838–40 (C730); Inventory and Appraisement Book, 1835–40 (C730); Inventory, Appraisement, and Sale Book, 1854–58 (C732), South Carolina Department of Archives and History, Columbia.

81. Record of the Office of the Probate Judge, Spartanburg County: Inventories, 1787–1810; Inventory and Appraisement Book, 1811–18 (C729); Inventory and Appraisement Book, 1818–25 (C730); Inventory and Appraisement Book, 1830–35 (C730); Inventory and Appraisement Book, 1838–40 (C730); Inventory and Appraisement Book, 1835–40 (C730); Inventory, Appraisement, and Sale Book, 1854–58 (C732), South Carolina Department of Archives and History, Columbia. The meaning of the distinction between mirrors and looking glasses here is unknown, although it may reflect the reentry of the term *mirror* into common use around 1860. See David Barquist, *American Tables and Looking Glasses in the Mabel Brady Garvan and Other Collections at Yale University* (New Haven, CT: Yale University Art Gallery, 1992), 294.

82. See David Maldwyn Ellis, *New York: State and City* (Ithaca, NY: Cornell University Press, 1979), 183; and Martin Bruegel, *Farm, Shop, Landing: The Rise of a Market Society in the Hudson Valley, 1780–1860* (Durham, NC: Duke University Press, 2002), 16, 90.

83. Bruegel, *Farm, Shop, Landing*, 161. Although I traveled to the Vedder Research Library in Coxsackie, New York, which houses the Greene County records, and conducted probate record research there, a reorganization of the records made it difficult to achieve a sufficient sample size, so here I rely on Bruegel's earlier work, which is based on a study of ninety-six probate records from Greene County. For this reason I do not have quartile data for Greene County. The slow rate of increase in mirror ownership over the first half of the nineteenth century may be explained in part by the significant number of settlers who were already well settled in the county at the time of its designation as such.

84. See Elizabeth A. Perkins, "The Consumer Frontier: Household Consumption in Early Kentucky," *Journal of American History* 78, no. 2 (1991): 488; Lowell Harrison and James Klotter, *A New History of Kentucky* (Lexington: University Press of Kentucky, 1997), 99; and Lewis Collins, *History of Kentucky* (1874; repr., Frankfort: Kentucky Historical Society, 1966), 2:169–229, 266, 268. See also Allen J. Share, *Cities in the Commonwealth: Two Centuries of Urban Life in Kentucky* (Lexington: University Press of Kentucky, 1982), 1–31.

85. Perkins, "The Consumer Frontier," 488–89; Fayette County Will Books: A–C (microfilm roll no. 987393); K–M (microfilm roll no. 987396); S–U (microfilm roll no. 987399), Kentucky Historical Society, Frankfort.

86. Moreover, the probate records only reveal the minimum number of mirrors in any of these nineteenth-century households. As Holly Izard has observed, it was possible that the goods of one person who lived in the house (e.g., a wife) that were not considered to be owned by the decedent might not be included in the inventory. See Izard, "Random or Systematic?" 153–54. It is also possible that some small pocket mirrors may have, from time to time, been present in a household—perhaps even on the body of a person not home when the assessors paid their visit or tucked away—but not counted in the estate inventory.

87. Charles Dickens, "Our Mutual Friend," *Harper's New Monthly Magazine* 29, June 1864, 84.

CHAPTER 4: Reliable Mirrors and Troubling Visions

1. Richard Devens, *The Pictorial Book of Anecdotes and Incidents of the War of the Rebellion, Civil, Military, Naval, and Domestic . . .* (Hartford, CT: Hartford Publishing, 1866), 648, https://archive.org/details/pictorialbookofa03deve. Versions of this story also appear in Henry J. Raymond, *The Life and Public Services of Abraham Lincoln* (New York: Darby and Miller, 1865), 750–51; and F. B. Carpenter, *Six Months at the White House with Abraham Lincoln* (New York: Hurd and Houghton, 1866), 163–65.

2. Lincoln's shaving mirror, for example, is housed in the collection of the Abraham Lincoln Presidential Library and Museum in Springfield, Illinois (accession no. LR 149). The mirror is 13 in. wide by 17 in. tall by 7.5 in. deep. It was likely made and purchased between 1849 and 1860. Email correspondence with James Cornelius, Curator, Lincoln Collection, July 2016.

3. Devens, *The Pictorial Book of Anecdotes*, 648; Jonathan Crary, *Techniques of the Observer: On Vision and Modernity in the Nineteenth Century* (Cambridge, MA: MIT Press, 1992), 97–136.

4. Erving Goffman, *The Presentation of Self in Everyday Life* (New York: Anchor, 1959), 4, 23–24. Although Goffman is sensitive to the ways in which men and women formulate their public selves, he does not consider the role of the mirror in the construction of the personal front.

5. "My Aunt Dagon's Mission," *Harper's Weekly*, Sept. 2, 1865, 554. *Harper's Weekly* was "the definitive newspaper of record" in the United States, with a subscription base, at various

times, of between one hundred thousand and three hundred thousand and an "effective readership of at least half a million people." See *Harper's Weekly, 1857–1912*, Alexander Street, http://alexanderstreet.com/products/harpers-weekly-1857-1912.

6. See Eliza Haywood, *Selections from the Female Spectator*, ed. Patricia Meyer Spacks (1744–46; repr., New York: Oxford University Press, 1999), 68. On the male gaze see Laura Mulvey, *Visual and Other Pleasures*, 2nd ed. (New York: Palgrave Macmillan, 2009), 19–27.

7. Anne Home (Nancy) Shippen Livingston, *Nancy Shippen: Her Journal Book . . .*, comp. and ed. Ethel Armes (Philadelphia: J. B. Lippincott, 1935), 142, https://archive.org/details/nancyshippenherj006968mbp; Eleanor Lewis, "Letter from Eleanor Parke Custis Lewis to Elizabeth Bordley, October 14, 1822," in *George Washington's Beautiful Nelly: The Letters of Eleanor Parke Custis Lewis to Elizabeth Bordley Gibson, 1794–1851*, ed. Patricia Brady (Columbia: University of South Carolina Press, 1991), 128, originally accessed at North American Women's Letters and Diaries (hereafter NAWLD), a subscription-based digital archive available through Alexander Street Press and located at http://alexanderstreet.com/products/north-american-womens-letters-and-diaries; Elizabeth Blackwell, *Pioneer Work in Opening the Medical Profession to Women: Autobiographical Sketches by Dr. Elizabeth Blackwell* (London: Longmans, Green, 1895), originally accessed at NAWLD; Annie Van Ness, *Diary of Annie L. Van Ness, 1864–1881* (Alexandria, VA: Alexander Street Press, 2004), 99, NAWLD.

8. Lucy Larcom, *A New England Girlhood* (New York: Corinth Books, 1961), 106; "Civilization," *Harper's New Monthly Magazine*, Jan. 1858, 177.

9. Louisa May Alcott, *Louisa May Alcott: Her Life, Letters, and Journal*, ed. Ednah D. Cheney (Boston: Roberts Bros., 1889), 60, https://archive.org/details/louisamay00alcorich, originally accessed at NAWLD; Abigail Adams, *Adams Family Correspondence*, vol. 5, ed. Richard Alan Ryerson et al. (Cambridge, MA: Harvard University Press, 1992), 433–35; Larcom, *A New England Girlhood*, 106.

10. Abigail Adams, *Letters of Mrs. Adams*, vol. 2, ed. Charles Francis Adams (Boston: Little, Brown, 1841), 264; Esther Hill Hawks, *A Woman Doctor's Civil War*, ed. Gerald Schwartz (Columbia: University of South Carolina Press, 1984), 1, 91; Alcott, *Louisa May Alcott*, 247.

11. Frances E. Willard, *Nineteen Beautiful Years; or, Sketches of a Girl's Life* (New York: Harper and Brothers, 1864), 147; Eliza Frances Andrews, *The War-Time Journal of a Georgia Girl, 1864–1865*, ed. Spencer B. King Jr. (New York: Appleton-Century-Crofts, 1908), 298–99, originally accessed at NAWLD. See chapter 5 for other ways that white women attempted to make this work seem effortless.

12. Sarah Morgan, *The Civil War Diary of a Southern Woman* (New York: Simon and Schuster, 1991), 80, 521, 528. Morgan sat down at her dressing table to record her first diary entry, in what would grow to be an expansive record of her Civil War–era experiences, at the beginning of 1862. Like most white women in the mid-nineteenth century, Morgan was intimately familiar with her mirror self. As a member of a wealthy family, she had grown up in a material environment overflowing with mirrors in the public and private spaces of her home.

13. Swinging mirrors were advertised as early as 1711 in Philadelphia and 1737 in Boston. See Elisabeth Donaghy Garrett, "Looking Glasses in America: 1700–1850," in David L. Barquist, *American Tables and Looking Glasses in the Mabel Brady Garvan and Other Collections at Yale University* (New Haven, CT: Yale University Press, 1992), 28–30; and Benno M. Forman, "Furniture for Dressing in Early America, 1650–1730: Forms, Nomenclature, and Use," *Winterthur Portfolio* 22, no. 2/3 (1987): 157.

14. Eliza Leslie, *The House Book: or, A Manual of Domestic Economy* (1840; repr., Philadelphia: Carey and Hart, 1845), 300.

15. Leslie, *The House Book*, 300–301; Eliza Leslie, *Miss Leslie's Behaviour Book: A Guide and Manual for Ladies* (1849; repr., New York: Arno, 1972), 31–32; Van Ness, *Diary of Annie Van Ness*, 530; "Toilet Glass" advertisement from the firm of Chappell and Godden, *Harper's Weekly*, June 16, 1866, 382.

16. Leslie, *Miss Leslie's Behaviour Book*, 92; "Diary of a Refugee Summary," http://docsouth.unc.edu/fpn/fearn/summary.html; Frances Fearn, ed., *Diary of a Refugee* (New York: Moffat, Yard, 1910), 58, http://docsouth.unc.edu/fpn/fearn/menu.html; "Comstock, Elizabeth Leslie Rous Wright, 1815–1891," NAWLD; Elizabeth Comstock, *Life and Letters of Elizabeth L. Comstock* (Philadelphia: John C. Winston, 1895), 152, originally accessed at NAWLD.

17. Louise Clappe, *The Shirley Letters: From the California Mines in 1851–1852*, ed. Thomas C. Russell (San Francisco: Thomas C. Russell, 1922), 98, 333, originally accessed at NAWLD; Kenneth L. Holmes, ed., *Covered Wagon Women: Diaries and Letters from the Western Trails*, vol. 2, *1850* (Lincoln: University of Nebraska Press, 1996), 51.

18. John A. H. Sweeney, ed., "The Norris-Fisher Correspondence: A Circle of Friends, 1779–1782," *Delaware History* 6 (1954–55): 199–200; Lucy Breckinridge, *Lucy Breckinridge of Grove Hill: The Journal of a Virginia Girl, 1862–1864*, ed. Mary D. Robertson (Columbia: University of South Carolina Press, 1994), 214–15.

19. Oliver Wendell Holmes, "The Stereoscope and the Stereograph," *Atlantic*, June 1859, www.theatlantic.com/magazine/archive/1859/06/the-stereoscope-and-the-stereograph/303361/.

20. Jane Adlin, "Vanities: Art of the Dressing Table," *Metropolitan Museum of Art Bulletin* 71, no. 2 (2013): 30–33; Alexandra Parker, "Reflections after the Fire: The History of the Monroe Shaving Mirrors," *White House History* 35 (Summer 2014): 103–5.

21. Elisha Mitchell, "Letter from Elisha Mitchell to Maria North, February 11, 1818," http://docsouth.unc.edu/true/mss01-18/mss01-18.html, 3; Lucius Fairchild, *California Letters of Lucius Fairchild* (Madison: State Historical Society of Wisconsin, 1931), 38.

22. This powder horn is in the private collection of Rich Nardi and can be viewed at http://americanpowderhorns.com/?p=244. My thanks to Mr. Nardi for providing me very helpful information about the powder horn in an email exchange (Oct. 2014). An interesting variant, also in the Nardi collection, is a powder horn with a piece of clear glass as the base plug, under which has been placed a slip of paper bearing the powder horn owner's name (Ezekiel Kelley) and the date (1763). It can be viewed at http://americanpowderhorns.com/?p=237.

23. George C. Neumann and Frank J. Kravic include examples of Revolutionary War–era mirrors in *Collector's Illustrated Encyclopedia of the American Revolution* (Harrisburg, PA: Stackpole Books, 1975), 241.

24. See Carlton McCarthy, *Detailed Minutiae of Soldier Life in the Army of Northern Virginia, 1861–1865* (1882; repr., Lincoln: University of Nebraska Press, 1993), 17.

25. Henry Warren Howe, *Passages from the Life of Henry Warren Howe, Consisting of Diary and Letters Written during the Civil War, 1861–1865* (Lowell, MA: Courier-Citizen Co., Printers, 1899), 131, originally accessed at The American Civil War: Letters and Diaries (hereafter ACWLD), a subscription-based digital archive available through Alexander Street Press and located at http://alexanderstreet.com/products/american-civil-war-letters-and-diaries; Hallock Armstrong, *Letters from a Pennsylvania Chaplain at the Siege of Petersburg, 1865*, ed. Hallock F. Raup (privately published, 1961), 34, ACWLD.

26. Louis Leon, *Diary of a Tar Heel Confederate Soldier* (Charlotte, NC: Stone Publishing, 1913), 1, http://docsouth.unc.edu/fpn/leon/leon.html; William Thompson Lusk, *War Letters of William Thompson Lusk, Captain, Assistant Adjutant-General, United States Volunteers, 1861–1863* (New York: privately printed, 1911), 83, originally accessed at ACWLD.

27. Osborn H. Oldroyd, *A Soldier's Story of the Siege of Vicksburg, from the Diary of Osborn H. Oldroyd* (Springfield, IL: privately published, 1885), 16–17, originally accessed at ACWLD. Oldroyd kept the "little mirror that had helped him make his last toilet" as a remembrance of his friend.

28. Emily E. Judson, *Memoir of Sarah B. Judson, Member of the American Mission to Burmah* (Cincinnati, OH: L. Colby, 1849), 130, NAWLD; Susan Dabney Smedes, *Memorials of a Southern Planter* (Baltimore: Cushings and Bailey, 1887), 168, http://docsouth.unc.edu/fpn/smedes/smedes.html; James Henry Hammond, *Secret and Sacred: The Diaries of James Henry Hammond, a Southern Slaveholder*, ed. Carol Bleser (Columbia: University of South Carolina Press, 1997), 300.

29. Phoebe Yates Pember, *A Southern Woman's Story* (New York: G. W. Carleton, 1879), 107–9, originally accessed at NAWLD.

30. John K. Townsend, "Narrative of Journey across the Rocky Mountains, to the Columbia River," in *Early Western Travels: 1748–1846*, vol. 21, ed. Reuben Gold Thwaites (Cleveland, OH: A. H. Clark, 1905), 277.

31. Peter Edes, *Horrid Massacre!! Sketches of the Life of Captain James Purrinton Who on the Night of the Eighth of July, 1806, Murdered His Wife, Six Children, and Himself* (Augusta, ME: Peter Edes, 1806), 7; George Templeton Strong, *The Diary of George Templeton Strong: Young Man in New York, 1835–1849*, vol. 1, ed. Allan Nevins and Milton Halsey Thomas (New York: Macmillan, 1952), 152; "Suicide after Smallpox," *New York Times*, May 2, 1902.

32. As scientific instruments, beginning in the eighteenth century, mirrors began to "make a vital contribution to our understanding of the universe." Benjamin Goldberg, *The Mirror and Man* (Charlottesville: University Press of Virginia, 1985), 175, 179–239. I have chosen to focus here, however, on how men used their own mirrors, in which they encountered their mirror selves, to extend their vision beyond their own faces and bodies.

33. Paul F. Mottelay and T. Campbell-Copeland, eds., *The Soldier in Our Civil War: A Pictorial History of the Conflict, 1861–1865*, vol. 2 (New York: Stanley Bradley Publishing, 1893), 280.

34. Samuel R. Watkins, *"Co. Aytch:" Maury Grays First Tennessee Regiment; or, A Side Show of the Big Show* (Chattanooga, TN: Times Printing, 1900), 142, originally accessed at ACWLD; Oldroyd, *A Soldier's Story*, 47–49.

35. Samuel Groome, "A Glass for the People of New-England" (London: n.p., 1676), http://quod.lib.umich.edu/cgi/t/text/text-idx?c=eebo2;idno=A42186.0001.001; Lewis Bayly, *Meditations and Prayers for Household Piety: Taken Out of the Practice of Piety, to Which Is Added, "The Drunkard's Looking-Glass: or, A Short View of Their Present Shame and Future Misery"* (Boston: Printed for Alford Butler, 1728); George Churchman, "A Little Looking-Glass for the Times; or, A Brief Remembrancer" (Wilmington, DE: James Adams, 1764), iii, http://digitallibrary.hsp.org/index.php/Detail/Object/Show/object_id/13722; John Adams to Abigail Adams, Nov. 15, 1800, www.masshist.org/digitaladams/archive/doc?id=L18001115ja; "The Mirror of Misery; or, Tyranny Exposed" (New York: Samuel Wood, 1807), 8.

36. The only woman I have found who wrote about refusing to look into mirrors was the eighteenth-century Lady Mary Wortley Montagu from England, whose face had been severely disfigured by smallpox. See Lady Mary Wortley Montagu, *The Letters and Works of Lady Mary Wortley Montagu*, vol. 3 (London: Richard Bentley, 1837), 171.

37. "Law and Lawyers," *Southern Quarterly Review* 6, no. 12 (1844): 424, http://name.umdl.umich.edu/acp1141.1-06.012; James S. Pike, *First Blows of the Civil War: The Ten Years of Preliminary Conflict in the United States, from 1850–1860* (New York: American News Company, 1879), 305, https://archive.org/details/firstblowsofcivioopike.

38. See Jon Butler, *Awash in a Sea of Faith: Christianizing the American People* (Cambridge, MA: Harvard University Press, 1990), 67–97.

39. John Hale, *A Modest Enquiry into the Nature of Witchcraft* (Boston: B. Green and J. Allen, 1702), 134–35, http://name.umdl.umich.edu/N00872.0001.001.

40. Lydia Sexton, *Autobiography of Lydia Sexton: The Story of Her Life through a Period over Seventy-Two Years, from 1799–1872* (1882; repr., New York: Garland, 1987), 90–91.

41. "May Day: A Village Tale Founded on Fact," *Southern Literary Messenger* 10 (Sept. 1844): 548–50.

42. *Harper's Weekly*, March 9, 1867, 158.

43. I suspect that cloth was more commonly used than the practice of turning the mirror around to face the wall. Turning mirrors around would have increased the likelihood that valuable, large wall mirrors would have fallen off the wall and broken, either in the act of turning or by loosening the hold of the fastener to the wall.

44. James Frazer, *The Golden Bough: A Study in Magic and Religion, Part II: Taboo and the Perils of the Soul* (London: Macmillan, 1914), 94–95. All of Frazer's evidence was from after the advent of accurately reflective glass mirrors. Covering looking glasses at the time of death was also practiced in Ireland into the twentieth century, according to Elaine Ní Bhraonáin, author of "Aspects of the Irish Way of Death: Materials and Customs of the Wake" (bachelor's thesis, University College Dublin, 2002), email correspondence with the author, 2005. Another group who covered mirrors after a death were people of the Jewish faith. In *The Jewish Book of Why*, in addition to repeating the reasons given by Frazer, Alfred Kolatch adds:

> The most popular explanation is that mirrors are associated with personal vanity. During a period of mourning, it is not appropriate to be concerned with one's personal appearance. The covering of mirrors has also been explained as an expression of the mourner's belief that despite the great loss just suffered, he refuses to blame God. To see himself in a sorry state, as a grieving mourner, is not a compliment to God, since man was created in the image of God. Another reason given for the covering of mirrors is that prayer services are held in the house of mourning, and it is forbidden to pray in front of a mirror. (Synagogues are not decorated with mirrors.) (Kolatch, *The Jewish Book of Why* [Middle Village, NY: Jonathan David, 1981], 64–65.)

45. For a detailed discussion of the Punderson embroidery see Laurel Thatcher Ulrich, *The Age of Homespun: Objects and Stories in the Creation of an American Myth* (New York: Knopf, 2001), 228–47.

46. Cortlandt Van Rensselaer, *Funeral Sermon, . . . after the Decease of William Henry Harrison . . .* (Washington, 1841), 49; William T. Coggeshall, *The Journeys of Abraham Lincoln from Springfield to Washington . . .* (Columbus: Ohio State Journal, 1865), 110, https://archive.org/details/lincolnmemorialjooincogg.

47. "Aunt Clussey," in *The American Slave: A Composite Autobiography*, ed. George P. Rawick, supp. ser. 1, vol. 1, *Alabama Narratives* (Westport, CT: Greenwood, 1977), 20.

48. Lloyd C. Douglas, *Time to Remember* (Boston: Houghton Mifflin, 1951), 222, https://archive.org/details/timetorememberoodoug.

49. In *Mirror, Mirror: A History of the Human Love Affair with Reflection*, Mark Pendergrast traces the belief to the Romans, who "believed that breaking a mirror would

cause seven years' bad luck (they thought a person's health changed in seven-year cycles" (New York: Basic Books, 2003), 30. For African Americans discussing this belief, see chapter 5.

50. James Hall, *Letters from the West: Containing Sketches of Scenery, Manners, and Customs; and Anecdotes Connected with the First Settlements of the Western Sections of the United States* (London: Henry Colburn, 1828), 329–30, 339, https://archive.org/details/lettersfrom westoohallgoog.

51. "Miscellany," *Harper's Weekly*, Jan. 16, 1858, 44. The text that appeared in this edition of *Harper's Weekly* can also be found forty-five years earlier in John Brand, *Observations on Popular Antiquities* . . . , vol. 2, ed. Henry Ellis (London: privately printed, 1813), 491–92, https://play.google.com/store/books/details?id=1B9JAAAAcAAJ&rdid=book -1B9JAAAAcAAJ&rdot=1.

52. Mrs. Burton Harrison, *Recollections Grave and Gay* (New York: Charles Scribner's Sons, 1911), 201–3, http://docsouth.unc.edu/fpn/harrison/menu.html.

53. "Small Superstitions," *Harper's Weekly*, August 7, 1858, 499–500.

54. Devens, *The Pictorial Book*, 648.

55. There is one provocative piece of evidence that beliefs in reflective power may have persisted into the twentieth century: the archaeological discovery of two mirrors that had been "intentionally concealed through human agency" on the Overturf homestead near Darby, Montana, in the early twentieth century. These large mirrors had been concealed behind a "specially constructed false wall in the attic." C. Riley Augé contends that "the complete inaccessibility of the mirrors suggests they were intentionally placed there with no objective for later retrieval." See C. Riley Augé, "Mirror to the Past: The Material Culture of Traditional Belief on an Early Twentieth Century Montana Homestead," 3, 16–17, https://umontana.academia.edu/CRileyAuge.

56. The idea that vision was the "noblest sense" had come from seventeenth-century French philosopher René Descartes, who wrote that "all the management of our lives depends on the senses, and since that of sight is the most comprehensive and the noblest of these, there is no doubt that the inventions which serve to augment its power are among the most useful that there can be." René Descartes, *Discourse on Method, Optics, Geometry, and Meteorology* (Indianapolis: Bobbs-Merrill, 1965), 65, quoted in Martin Jay, *Downcast Eyes: The Denigration of Vision in Twentieth-Century French Thought* (Berkeley: University of California Press, 1994), 71. Along with other developments, especially the development of print technology, mirrors contributed to the elevation of human sight above the other senses until sight became, renowned scholar of orality and literacy Walter Ong argued, "the only valid way of knowing the world." The argument about print technology was articulated by Marshall McLuhan and Walter Ong and is summarized by Martin Jay (*Downcast Eyes*, 66–67). Art historian Jonathan Crary explores this elevation of vision as the "noblest sense" in *Techniques of the Observer*, 25–66. For more on the human reliance on vision since at least the era of the Western European Renaissance see Joachim-Ernst Berendt, *The Third Ear: On Listening to the World*, trans. Tim Nevill (New York: Henry Holt, 1992), 21–23.

57. This paragraph briefly summarizes the arguments put forth by Jonathan Crary in chapter 3, "Subjective Vision and the Separation of the Senses," and chapter 4, "Techniques of the Observer," in *Techniques of the Observer*, 67–136.

58. Ellen Tucker Emerson, *The Letters of Ellen Tucker Emerson*, ed. Edith W. Gregg (Kent, OH: Kent State University Press, 1982), 1:32, NAWLD.

59. James Cook, *The Arts of Deception: Playing with Fraud in the Age of Barnum* (Cambridge, MA: Harvard University Press, 2001), 214–15; Peter Brownlee, "'The Economy of

the Eyes': Vision and the Cultural Production of Market Revolution, 1800–1860" (PhD diss., George Washington University, 2003).

60. Crary argues that the devices he studies—including the stereoscope—fell into disuse because people could not tolerate the questions they raised about whether what could be seen could be trusted to be an accurate representation of reality. As a result, Crary argues, photography, which did not raise these troubling questions, triumphed. See Crary, *Techniques of the Observer*, 133–36.

61. Alcott, *Louisa May Alcott*, 147.

62. Maria Lydig Daly, *Diary of a Union Lady, 1861–1865*, ed. Harold Earl Hammond (Lincoln: University of Nebraska Press, 2000), 84–85. It is also possible that Daly was looking into a mirror that produced a distortion because of an imperfection in the glass or coating. These were noted as late as 1872 by H. J. Rodgers, a photographer, who wrote that "we look differently in as many mirrors as we may choose to scrutinize. We therefore find it absolutely impossible to see ourselves as others see us. . . . Owing to the imperfections, and optical irregularities in many glasses, we are transformed into a complete abortive representation of the original." H. J. Rodgers, *Twenty-Three Years under a Sky-Light, or Life and Experiences of a Photographer* (Hartford, CT: H. J. Rodgers, 1872), 172–73.

63. Daniel Robinson Hundley, *Prison Echoes of the Great Rebellion* (New York: S. W. Green, 1874), 214, ACWLD.

64. James K. Hosmer, *The Color-Guard: Being a Corporal's Notes of Military Service in the Nineteenth Army Corps* (Boston: Walker, Wise, 1864), 149–50.

65. Morgan, *Civil War Diary*, 579.

66. Joseph P. Ferrie, "How Ya Gonna Keep 'Em Down on the Farm [When They've Seen Schenectady]? Rural-to-Urban Migration in 19th Century America, 1850–70," 4, http://faculty.wcas.northwestern.edu/~fe2r/papers/urban.pdf. See also Joseph P. Ferrie, "The End of American Exceptionalism? Mobility in the U.S. since 1850" (working paper 11324, Bureau of Economic Research, 2005), 17–18, www.nber.org/papers/w11324.

67. Karen Halttunen, *Confidence Men and Painted Women: A Study of Middle-Class Culture in America, 1830–1870* (New Haven, CT: Yale University Press, 1982), 36, 40.

68. Stephen W. Berry II, *All That Makes a Man: Love and Ambition in the Civil War South* (Oxford: Oxford University Press, 2003), 175–79; James M. Williams, *From That Terrible Field: Civil War Letters of James M. Williams, Twenty-First Alabama Infantry Volunteers*, ed. John Kent Folmar (Tuscaloosa: University of Alabama Press, 1981), 91.

69. Georgeanna Woolsey Bacon and Eliza Woolsey Howland, eds., *Letters of a Family during the War for the Union, 1861–1865* (privately published, 1899), 2:486, NAWLD.

70. "My First Patient," *Harper's Weekly*, April 29, 1865, 267.

CHAPTER 5: Fashioning Whiteness

1. Valerie Babb, *Whiteness Visible: The Meaning of Whiteness in American Literature and Culture* (New York: New York University Press, 1998), 43; Nancy Shoemaker, *A Strange Likeness: Becoming Red and White in Eighteenth-Century North America* (Oxford: Oxford University Press, 2004), 129.

2. Bridget T. Heneghan, *Whitewashing America: Material Culture and Race in the Antebellum Imagination* (Jackson: University Press of Mississippi, 2003), xxiv, 15.

3. See Karen Ordahl Kupperman, *Indians and English: Facing Off in Early America* (Ithaca, NY: Cornell University Press, 2000), 148–53, 150.

4. Reuben Gold Thwaites, ed., *Jesuit Relations and Allied Documents*, vol. 38 (Cleveland, OH: Burrows Brothers, 1901), 253; Stewart Rafert, *The Miami Indians of Indiana: A Persistent People, 1654–1994* (Indianapolis: Indiana Historical Society, 1996), 4.

5. Emma Helen Blair, trans. and ed., *The Indian Tribes of the Upper Mississippi Valley and Region of the Great Lakes* . . . , vol. 1 (1911; repr., Lincoln: University of Nebraska Press, 1996), 141–42.

6. See Josephine Paterek, *Encyclopedia of American Indian Costume* (New York: Norton, 1996), 53–55.

7. Pierre François Xavier de Charlevoix, *Journal of a Voyage to North-America*, vol. 2 (1761; repr., Chicago: Caxton Club, 1923), 37, https://archive.org/details/journalofvoyageto2char.

8. Peter Kalm, *Peter Kalm's Travels in North America: The English Version of 1770*, ed. Adolph B. Benson, vol. 2 (New York: Wilson-Erickson, 1937), 519–21.

9. Here again, although Gregg noted that women wore less paint and ornamentation, he does not actually discuss whether they used mirrors. Josiah Gregg, "The Commerce of the Prairies," in *Early Western Travels, 1748–1846*, vol. 20, ed. Reuben Gold Thwaites (Cleveland, OH: Arthur H. Clark, 1905), 329–30, https://books.google.com/books?id=M5k_AQAAMAAJ&; David Dary, "Gregg, Josiah (1806–1850)," *Encyclopedia of Oklahoma History and Culture*, www.okhistory.org/publications/enc/entry.php?entry=GR027.

10. Alexander Ross, *The Fur Hunters of the Far West*, ed. Kenneth A. Spaulding (1855; repr., Norman: University of Oklahoma Press, 1956), 155.

11. Gregg, *Commerce of the Prairies*, 330.

12. James Hobbs, *Wild Life in the Far West: Personal Adventures of a Border Mountain Man* (1872; repr., Glorieta, NM: Rio Grande Press, 1969), 39. On the importance of time consciousness to white men in antebellum America see Mark M. Smith, *Mastered by the Clock: Time, Slavery, and Freedom in the American South* (Chapel Hill: University of North Carolina Press, 1997), esp. 93–127.

13. "Notes on the Early Life of Chief Washakie, Taken Down by Captain Ray," ed. Don D. Fowler, *Annals of Wyoming* 36, no. 1 (1964): 39–40, https://archive.org/stream/annalsofwyom36121964wyom.

14. Nancy Hathaway, *Native American Portraits: 1862–1918, Photographs from the Collection of Kurt Koegler* (San Francisco: Chronicle, 1990), 29. See also Joanna Cohan Scherer, "You Can't Believe Your Eyes: Inaccuracies in Photographs of North American Indians," *Studies in the Anthropology of Visual Communication* 2, no. 2 (1975): 67–79.

15. That the threat posed by Native Americans had by then been obliterated is represented perhaps most clearly in the photograph of Gorosimp, where the placement of the gun is reminiscent of a flaccid penis.

16. Richard Hakluyt, *Divers Voyages Touching the Discovery of America and the Islands Adjacent*, ed. John Winter Jones (1582; repr., London: Hakluyt Society, 1850), 124; Roger Williams, *A Key into the Language of America*, in *The Complete Works of Roger Williams*, vol. 1, ed. J. H. Trumbull (1866; repr., New York: Russell and Russell, 1963), 184 (unbracketed page numbers cited).

17. Orlando M. McPherson, *Indians of North Carolina* . . . (Washington: Government Printing Office, 1915), 203, http://docsouth.unc.edu/nc/mcpherson/mcpherson.html; Meriwether Lewis and William Clark, *Original Journals of the Lewis and Clark Expedition, 1804–1806*, vol. 6, ed. Reuben Gold Thwaites (New York: Dodd, Mead, 1905), 273; George Catlin, *Episodes from Life among the Indians and Last Rambles*, ed. Marvin C. Ross (Norman: University of Oklahoma Press, 1959), 231; Gilbert Malcolm Sproat, *The Nootka: Scenes and Studies of Savage Life*, ed. Charles Lillard (Victoria, British Columbia: Sono Nis Press, 1987), 23.

18. Samuel W. Pond, *The Dakota or Sioux in Minnesota as They Were in 1834* (St. Paul: Minnesota Historical Society, 1986), vii–xi, 36.

19. Patricia Albers and Beatrice Medicine, *The Hidden Half: Studies of Plains Indian Women* (Lanham, MD: University Press of America, 1983), 1–5.

20. Although historians have not spent much time studying the incorporation of mirrors into Native American societies, when they have done so, they have perpetuated this idea that only Native American men used mirrors. Charles Hanson argued that the users of looking glasses were "principally men," while Wilbur Jacobs scoffed at the "vanity" of the "ridiculous warriors" who demanded mirrors in trading with European Americans. Even James Axtell's important analysis of the "first consumer revolution" in seventeenth-century Native American life assumes that only Native American men used looking glasses. See Charles E. Hanson Jr., "Trade Mirrors," *Museum of the Fur Trade Quarterly* 22, no. 4 (1986): 1; Wilbur R. Jacobs, *Wilderness Politics and Indian Gifts: The Northern Colonial Frontier, 1748–1763* (Lincoln: University of Nebraska Press, 1966), 78; and James Axtell, "The First Consumer Revolution," in *Consumer Society in American History: A Reader*, ed. Lawrence B. Glickman (Ithaca, NY: Cornell University Press, 1999), 93.

21. Richard Irving Dodge, *Our Wild Indians: Thirty-Three Years' Personal Experience among the Red Men of the Great West . . .* (Hartford, CT: A. D. Worthington, 1883), 431, https://archive.org/details/indiansourwildthoododgrich.

22. John McLean [*sic*], *The Indians of Canada: Their Manners and Customs* (London: Charles H. Kelly, 1892), 28; Susan Gray, "Maclean, John," in *Dictionary of Canadian Biography*, vol. 15, University of Toronto/Université Laval, 2003, www.biographi.ca/en/bio/maclean_john_15E.html.

23. Hobbs, *Wild Life in the Far West*, 39.

24. White arguments for Native assimilation into white society emerged, alongside removal, by the beginning of the nineteenth century and grew more prominent by the end of that century. As early as 1803, President Thomas Jefferson laid out a plan in a letter to William Henry Harrison, the territorial governor of Indiana, by which Native people "will in time either incorporate with us as citizens of the United States, or remove beyond the Mississippi." Thomas Jefferson, *Political Writings*, ed. Joyce Appleby and Terence Ball (Cambridge: Cambridge University Press, 1999), 525. See also Bernard W. Sheehan, *Seeds of Extinction: Jeffersonian Philanthropy and the American Indian* (Chapel Hill: University of North Carolina Press, 1973); and Jeffrey Ostler, *The Plains Sioux and U.S. Colonialism from Lewis and Clark to Wounded Knee* (New York: Cambridge University Press, 2004).

25. Henry Rowe Schoolcraft, *History of the Indian Tribes of the United States . . .*, vol. 6 (Philadelphia: Lippincott, 1857), 566. On the parlor as a feminine space see Daphne Spain, who identifies the nineteenth-century parlor as "a woman's space also open to outsiders," in *Gendered Spaces* (Chapel Hill: University of North Carolina Press, 1992), 123. For more on nineteenth-century parlors see Katherine C. Grier, *Culture and Comfort: Parlor Making and Middle-Class Identity, 1850–1930* (Washington: Smithsonian Institution Press, 1988).

26. Henry Rowe Schoolcraft, *Information Respecting the History, Condition and Prospects of the Indian Tribes of the United States . . .*, vol. 3 (Philadelphia: Lippincott, Grambo, 1853), 535.

27. Médéric-Louis-Elie Moreau de Saint-Méry, *A Civilization That Perished: The Last Years of White Colonial Rule in Haiti*, ed. and trans. Ivor D. Spencer (Lanham, MD: University Press of America, 1985,) 47.

28. Frances Anne Kemble, *Journal of a Residence on a Georgia Plantation in 1838–1839* (New York: Harper and Brothers, 1863), 261.

29. There have been criticisms of the WPA interviews. Because of the distance of seventy-five years or more between the interview and the events being remembered, a clear picture of the past may have been fundamentally distorted or irretrievable. There are also

serious concerns about the relationship between the interviewers, who were often white, and the African American interviewees: white interviewers may have tried to craft the testimony to present a more positive view of slavery, and there were simply things African Americans would not say in front of whites. Moreover, in light of the difficulty of the Great Depression these elderly men and women were enduring in the 1930s, interviewees may have remembered their childhood even under slavery fondly. See Norman R. Yetman, "The Background of the Slave Narrative Collection," *American Quarterly* 19, no. 3 (1967): 534–53; Norman R. Yetman, "Ex-Slave Interviews and the Historiography of Slavery," *American Quarterly* 36, no. 2 (1984): 181–210; Charles Joyner, *Down by the Riverside: A South Carolina Slave Community* (Urbana: University of Illinois Press, 1984), xv; and Emily West, *Chains of Love: Slave Couples in Antebellum South Carolina* (Urbana: University of Illinois Press, 2004), 5–8.

30. See George P. Rawick, *The American Slave: A Composite Autobiography*, vol. 1, *From Sundown to Sunup: The Making of the Black Community* (Westport, CT: Greenwood, 1972), 174–76; Emily West, *Chains of Love*, 5.

31. John Michael Vlach, *By the Work of Their Hands: Studies in Afro-American Folklife* (Ann Arbor: UMI Research Press, 1991), 66.

32. Charlie Davis interview, *Slave Narratives: A Folk History of Slavery in the United States from Interviews with Former Slaves* (hereafter *SN*), vol. 14, *South Carolina Narratives, Part 1* (Washington: Federal Writers' Project, 1941), 250. The *Slave Narratives* collection is available at the Library of Congress website: www.loc.gov/collections/slave-narratives-from-the-federal-writers-project-1936-to-1938/about-this-collection/.

33. "Bob Mobley," in *The American Slave: A Composite Autobiography*, ed. George P. Rawick, supp. ser. 1, vol. 4, pt. 2, *Georgia Narratives* (Westport, CT: Greenwood, 1977), 447.

34. Mary Frances Armstrong, Helen W. Ludlow, and Thomas P. Fenner, *Hampton and Its Students: By Two of Its Teachers, Mrs. M. F. Armstrong and Helen W. Ludlow. With Fifty Cabin and Plantation Songs, Arranged by Thomas P. Fenner* (New York: G. P. Putnam's Sons, 1874), 142–43, http://docsouth.unc.edu/church/armstrong/armstrong.html; Brent Kinser, "Summary of *Hampton and Its Students*," http://docsouth.unc.edu/church/armstrong/summary.html.

35. E. J. R. David and Annie O. Derthick, "What Is Internalized Oppression, and So What?" in *Internalized Oppression: The Psychology of Marginalized Groups*, ed. E. J. R. David (New York: Springer, 2014), 8–9.

36. George Yancy, *Black Bodies, White Gazes: The Continuing Significance of Race* (Lanham, MD: Rowman and Littlefield, 2008).

37. Joyner, *Down by the Riverside*, xxi, 90.

38. W. E. B. Du Bois, *The Souls of Black Folk* (Chicago: A. C. McClure, 1903), www.gutenberg.org/files/408/408-h/408-h.htm.

39. Cureton Milling interview, *SN*, vol. 14, *South Carolina Narratives, Part 3*, 196.

40. Henry Clay Bruce, *The New Man: Twenty-Nine Years a Slave, Twenty-Nine Years a Free Man* (York, PA: P. Anstadt and Sons, 1895), 130–31, http://docsouth.unc.edu/fpn/bruce/bruce.html.

41. William O'Neal, *Life and History of William O'Neal; or, The Man Who Sold His Wife* (St. Louis, MO: A. R. Fleming, 1896), 7, 10, 13–14, http://docsouth.unc.edu/neh/oneal/oneal.html.

42. Alexander Robertson interview, *SN*, vol. 14, *South Carolina Narratives, Part 4*, 32.

43. See Allyson Hobbs, *A Chosen Exile: A History of Racial Passing in American Life* (Cambridge, MA: Harvard University Press, 2014), 33–34.

44. See, e.g., Walter Johnson, "The Slave Trader, the White Slave, and the Politics of Racial Determination in the 1850s," *Journal of American History* 87, no. 1 (2000): 13–38; and Alan Trachtenberg, *Reading American Photographs: Images as History, Mathew Brady to Walker Evans* (New York: Hill and Wang, 1989), 53–60.

45. Mary Cathryn Cain, "The Art and Politics of Looking White: Beauty Practice among White Women in Antebellum America," *Winterthur Portfolio* 42, no. 1 (2008): 43.

46. Louise Clappe, *The Shirley Letters: From the California Mines in 1851–1852*, ed. Thomas C. Russell (San Francisco: Thomas C. Russell, 1922), 333, https://archive.org/stream/shirleylettersfoorussgoog#page/n12/mode/2up; Mollie Dorsey Sanford, *Mollie: The Journal of Mollie Dorsey Sanford in Nebraska and Colorado Territories, 1857–1866* (Lincoln: University of Nebraska Press, 1959), 143.

47. Cain, "The Art and Politics of Looking White," 43, 45. Kathy Peiss argues that cosmetics (makeup that did more than just highlight whiteness) gained acceptability among white women at the photographer's studio in the late nineteenth century. See Kathy Peiss, *Hope in a Jar: The Making of America's Beauty Culture* (New York: Henry Holt, 1999), 47.

CHAPTER 6: Mirrors in Black and Red

1. For a discussion of the penetration and circulation of mirrors in these areas see chapter 3.

2. Of the 109,000 captive Africans from West Central Africa who were taken to North America, 74,000 were transported to South Carolina. See David Eltis and David Richardson, *Atlas of the Transatlantic Slave Trade* (New Haven, CT: Yale University Press, 2010), map 137 (205), map 144 (216); and Christopher C. Fennell, "Group Identity, Individual Creativity, and Symbolic Generation in a BaKongo Diaspora," *International Journal of Historical Archaeology* 7, no. 1 (2003): 4–5. See chapter 3 for a discussion of the number of West Central Africans enslaved in North America.

3. The BaKongo belief system has been traced back to well before European contact in the late fifteenth century. See, e.g., Christopher C. Fennell, "BaKongo Identity and Symbolic Expression in the Americas," in *Archaeology of Atlantic Africa and the African Diaspora*, ed. Akinwumi Ogundiran and Toyin Falola (Bloomington: Indiana University Press, 2007), 203; and Robert Farris Thompson and Joseph Cornet, *Four Moments of the Sun: Kongo Art in Two Worlds* (Washington: National Gallery of Art, 1981), 28. The following discussion focuses on the subset of BaKongo beliefs specifically related to reflection. For a more comprehensive look at the BaKongo belief system, consult the authors cited throughout.

4. Wyatt MacGaffey, *Religion and Society in Central Africa: The BaKongo of Lower Zaire* (Chicago: University of Chicago Press, 1986), 43; Anita Jacobson-Widding, "The Encounter in the Water Mirror," in *Body and Space: Symbolic Models of Unity and Division in African Cosmology and Experience*, ed. Anita Jacobson-Widding (Stockholm: Almqvist and Wiksell International, 1991), 177–78.

5. Fennell, "BaKongo Identity and Symbolic Expression," 212; J. H. Oldam and G. A. Gollock, eds., *The International Review of Missions*, vol. 11 (London: Morrison and Gibb, 1922), 331; G. Cyril Claridge, *Wild Bush Tribes of Tropical Africa* (London: Seeley, Service, 1922), 275–76.

6. The belief systems of the Mbundu have been traced back to as early as the 1600s. See Robert W. Slenes, "The Great Porpoise-Skull Strike: Central African Water Spirits and Slave Identity in Early-Nineteenth-Century Rio de Janeiro," in *Central Africans and Cultural Transformations in the American Diaspora*, ed. Linda M. Heywood (Cambridge: Cambridge University Press, 2002), 192, 196.

7. Robert W. Slenes explores the Mbundu belief in *kianda* in his masterful analysis of what he calls the "Great Porpoise-Skull Strike," which took place near Rio de Janeiro in 1816 on a boat carrying the Englishman John Luccock. The boat's captain was a Portuguese sailor who had a crew of four enslaved Africans. Luccock, an "amateur naturalist," brought on board one day the skull of a porpoise. The Africans protested so vociferously to the skull's placement on the boat that the white men eventually threw the "obnoxious skull" overboard. Slenes argues that the enslaved crewmen took the porpoise skull to be a physical manifestation of one of these water spirits and believed it highly dangerous to remove such a spirit from the water. Surveying the landscape around which this incident took place, Slenes finds several physical signs that would have led the Africans to conclude that the porpoise skull was indeed imbued with such power, including several massive rock formations jutting out of the water. Slenes, "The Great Porpoise-Skull Strike," 183–201, esp. 184, 192, 196, 200.

8. Fennell, "BaKongo Identity and Symbolic Expression," 212–13; Fennell, "Group Identity," 13; Wyatt MacGaffey, *Art and Healing of the Bakongo Commented by Themselves: Minkisi from the Laman Collection* (Bloomington: Indiana University Press, 1991), 5; Wyatt MacGaffey, "Complexity, Astonishment and Power: The Visual Vocabulary of Kongo Minkisi," *Journal of Southern African Studies* 14, no. 2 (1988): 191, 201–2; Thompson and Cornet, *Four Moments of the Sun*, 28.

9. The texts were written by Laman's "Congolese research associates . . . young men, recently introduced to Christianity, literacy and the conventions of European narrative discourse, writing from personal experience that usually included field research." MacGaffey, "Complexity, Astonishment and Power," 188–89.

10. Karl Laman, *The Kongo III* (Lund, Sweden: Håkan Ohlssons Boktryckeri, 1962), 74; MacGaffey, *Art and Healing of the Bakongo*, 71–72 (brackets in original quote).

11. Laman, *The Kongo III*, 74. There is no evidence that metal mirrors were used in the making of minkisi.

12. Robert Farris Thompson, *Flash of the Spirit: African and Afro-American Art and Philosophy* (New York: Vintage, 1984); Laman, *The Kongo III*, 74; Thompson, *Four Moments of the Sun*, 198, 209n173. The definitions in brackets, added by the author for clarity, are from Jacobson-Widding, "The Encounter in the Water Mirror," 192, 203.

13. *Mbongo Nsimba* is held in the collection of Sweden's Museum of Mediterranean and Near Eastern Antiquities. This nkisi can be viewed at http://collections.smvk.se/carlotta-em/web/object/1063552. Wooden-figure minkisi can also have small mirror pieces for eyes (as *Mbongo Nsimba* does) and mirrors embedded in headdresses (as is the case in one of the two possible minkisi in the Pitt Rivers Museum [1884.65.65.1-.2]).

14. MacGaffey, "Complexity, Astonishment and Power," 199.

15. MacGaffey, *Art and Healing of the Bakongo*, 49. The Nkisi Matenzi is also held in the collection of Sweden's Museum of Mediterranean and Near Eastern Antiquities. It can be viewed at http://collections.smvk.se/carlotta-em/web/object/1188002. Because the nkisi is contained within a net bag, however, it is not visible in this photograph. Another nkisi in the collection for which the container was a (visible) snail shell, although without a mirror, can be seen at http://collections.smvk.se/carlotta-em/web/object/1191793. For more on minkisi see Wyatt MacGaffey, *Kongo Political Culture: The Conceptual Challenge of the Particular* (Bloomington: Indiana University Press, 2000), 78–133; Wyatt MacGaffey, *Astonishment and Power: The Eyes of Understanding Kongo Minkisi* (Washington: Smithsonian Institution, 1993); and Wyatt MacGaffey, "Fetishism Revisited: Kongo *Nkisi* in Sociological Perspective," *Africa: Journal of the International African Institute* 47 no. 2 (1977): 172–84.

16. Héli Chatelain, *Folk-Tales of Angola: Fifty Tales, with Ki-Mbundu Text, Literal English Translation, Introduction, and Notes* (1894; repr., New York: Negro Universities Press, 1969), 254; see also Gerald Moser, "Héli Chatelain: Pioneer of a National Language and Literature for Angola," *Research in African Literatures* 14, no. 4 (1983): 516–37.

17. Brigette C. Kamsler, "Finding Aid, MRL 1: Robert Hamill Nassau Papers, 1856–1976," Burke Library Archives at Union Theological Seminary, Columbia University Libraries, http://library.columbia.edu/content/dam/libraryweb/locations/burke/fa/mrl/ldpd_4492530.pdf; Robert Hamill Nassau, *Fetichism in West Africa: Forty Years' Observation of Native Customs and Superstitions* (New York: Charles Scribner's Sons, 1904), 108, 111–12, www.gutenberg.org/files/38038/38038-h/38038-h.htm.

18. Jacobson-Widding, "The Encounter in the Water Mirror," 208–11.

19. Of the total number of enslaved Africans held captive in North America, 13 percent originated from this region. See Fennell, "BaKongo Identity and Symbolic Expression," 203, 209.

20. Wilhelm Johann Müller, "Description of the Fetu Country, 1662–1669," in Adam Jones, *German Sources for West African History, 1599–1669* (Wiesbaden: Franz Steiner, 1983), 171–72. Efutu, the geographical center of the Fetu kingdom, was located approximately twelve miles inland from Cape Coast, in modern-day Ghana. See Holger Weiss, "The Entangled Spaces of Oddena, Oguaa and Osu: A Survey of Three Early Modern African Atlantic Towns, ca. 1650–1850," in *Ports of Globalisation, Places of Creolisation: Nordic Possessions in the Atlantic World during the Era of the Slave Trade*, ed. Holger Weiss (Leiden, The Netherlands: Brill, 2016), 34.

21. The nineteenth- and early twentieth-century collection practices of the Pitt Rivers collection make it impossible to trace with certainty the story attached to this mirror, but its use in divination seems possible given the other evidence of similar practices in the region. The original handwritten label on this object reads: "Éhun, mirror used in divination by Ahanta. This belongs to the Apowa fetish, of Betuah, a priestess of Apowa village, Ahanta, 16 miles from Sekondi, Gold Coast." Pres. by Capt. R. P. Wild, 1925. Accession no.: 1925.53.7.1–3. The record, but not an image, is available at http://objects.prm.ox.ac.uk/pages/PRMUID76668.html.

22. William Bosman, *A New and Accurate Description of the Coast of Guinea . . .* (London: Ballantyne Press, 1705), 158, https://archive.org/details/newaccuratedescr00bosm; Victor Turner, *The Ritual Process: Structure and Anti-structure* (1969; repr., New Brunswick, NJ: Aldine Transaction, 2008), 177–81; R. S. Rattray, *Ashanti* (1923; repr., New York: Negro Universities Press, 1969), 151, 163n2.

23. This area was a point of embarkation into slavery from both the Bight of Benin and the Bight of Biafra, which accounted for 16 percent and 12.7 percent of captives departing from Africa respectively. See Eltis and Richardson, *Atlas of the Transatlantic Slave Trade*, 14–15 (maps 8 and 9).

24. Richard and John Lander, *Journal of an Expedition to Explore the Course and Termination of the Niger*, vol. 2 (New York: Harper and Brothers, 1844), 177–78, https://archive.org/details/journalofexpedit02rich.

25. Samuel Crowther and John Christopher Taylor, *The Gospel on the Banks of the Niger: Journals and Notices of the Native Missionaries Accompanying the Niger Expedition of 1857–1859* (London: Dawsons of Pall Mall, 1968), vi, 251–52, 261.

26. "Reported Death of E. J. Glave," *Journal of the American Geographical Society of New York* 27 (1895): 187.

27. Edward James Glave, *In Savage Africa; or, Six Years of Adventure in Congo-Land* (New York: R. H. Russell and Son, 1892), 74.

28. John Michael Vlach, *The Afro-American Tradition in Decorative Arts* (Cleveland, OH: Cleveland Museum of Art, 1978), 143; Nassau, *Fetichism in West Africa*, 232.

29. Thompson, *Flash of the Spirit*, 104, 128–31; Leland Ferguson, *Uncommon Ground: Archaeology and Early African America, 1650–1800* (Washington: Smithsonian Books, 1992), 110–16; Mark P. Leone and Gladys-Marie Fry, with assistance from Timothy Ruppel, "Spirit Management among Americans of African Descent," in *Race and the Archaeology of Identity*, ed. Charles E. Orser Jr. (Salt Lake City: University of Utah Press, 2001), 146; Laura J. Galke, "Did the Gods of Africa Die? A Re-examination of a Carroll House Crystal Assemblage," *North American Archaeologist* 21, no. 1 (2000): 19–33; Amy L. Young, "Archaeological Evidence of African-Style Ritual and Healing Practices in the Upland South," *Tennessee Anthropologist* 21, no. 2 (1996): 139–55; Patricia M. Samford, *Subfloor Pits and the Archaeology of Slavery in Colonial Virginia* (Tuscaloosa: University of Alabama Press, 2007), 163–66; Fennell, "BaKongo Identity and Symbolic Expression," 221.

30. Samford, *Subfloor Pits*, 166; Fennell, "BaKongo Identity and Symbolic Expression," 221; K. L. Brown, "Material Culture and Community Structure: The Slave and Tenant Community at Levi Jordan's Plantation, 1848–1892," in *Working Toward Freedom: Slave Society and Domestic Economy in the American South*, ed. L. E. Hudson Jr. (New York: University of Rochester Press, 1994), 95–118. Scholars studying the Jordan Plantation argue that the material evidence suggests both the practice of BaKongo and Yoruban religions there (Fennell, "BaKongo Identity and Symbolic Expression," 223).

31. The story from the *Southern Literary Messenger* discussed in chapter 4 featured a young white girl who similarly saw an open grave when playing this game. "Charity Jones," in *The American Slave: A Composite Autobiography*, ed. George P. Rawick [hereafter *Am. Sl.*], supp. ser. 1, vol. 10, *Mississippi Narratives* (Westport, CT: Greenwood, 1977), 1201; Newbell Niles Puckett, *Folk Beliefs of the Southern Negro* (1926; repr., New York: Dover, 1969), 328, https://archive.org/details/folkbeliefsofsou00puck.

32. As Jeffrey E. Anderson describes it, conjuring "seeks to accomplish practical objectives through appeals to the spirit world" and is practiced by a conjurer, "a professional magic practitioner who typically receives payment in return for his or her goods and services." Jeffrey E. Anderson, *Conjure in African American Society* (Baton Rouge: Louisiana State University Press, 2005), x. On the importance of conjurers on southern plantations see Charles Joyner, *Shared Traditions: Southern History and Folk Culture* (Urbana: University of Illinois Press, 1999), 65.

33. Charles L. Perdue, Thomas E. Barden, and Robert K. Phillips, eds., *Weevils in the Wheat: Interviews with Virginia Ex-slaves* (Charlottesville: University Press of Virginia, 1976), 267–68.

34. The specific practices and meanings associated with the black cat bone varied by geographical location. See Puckett, *Folk Beliefs of the Southern Negro*, 257–58; Georgia Writers' Project, *Drums and Shadows: Survival Studies among the Georgia Coastal Negroes* (1940; repr., Westport, CT: Greenwood, 1973), 58, 102.

35. Thompson, *Flash of the Spirit*, 132; Vlach, *Afro-American Tradition*, 143.

36. Margaret Davis Cate and Orrin Sage Wightman, *Early Days of Coastal Georgia* (St. Simons Island, GA: Fort Frederica Association, 1955), 220–21; Georgia Writers' Project, *Drums and Shadows*, 116–17.

37. Thompson, *Four Moments of the Sun*, 199. This tradition has been documented at twentieth-century graves in Kongo, Georgia, and South Carolina. See Samford, "The Archaeology of African-American Slavery," 109.

38. In North America, African American beliefs and practices flourished alongside Native American and white practices. Moreover, there is significant evidence of overlap in these prac-

tices, suggesting perhaps both a universal human response to having access to reflection and the ways in which these three cultures in North America influenced one another.

39. Annie B. Boyd interview, *Slave Narratives: A Folk History of Slavery in the United States from Interviews with Former Slaves* (hereafter *SN*), vol. 7, *Kentucky Narratives* (Washington: Federal Writers' Project, 1941), 59. The *Slave Narratives* collection is available at the Library of Congress website: www.loc.gov/collections/slave-narratives-from-the-federal-writers-project-1936-to-1938/about-this-collection/; "William Emmons (Enslaved in Kentucky)," *Am. Sl.*, supp. ser. 1, vol. 5, *Ohio Narratives*, 328; Cheney Cross interview, *SN*, vol. 1, *Alabama Narratives*, 102. African Americans also reported similar beliefs to Newbell Niles Puckett. They "prescribe the placing of the fragments of the broken mirror in running water as a means of avoiding the ill luck (the trouble will pass away in seven hours), the running water being supposed to 'wash de trubbl' away.'" Puckett, *Folk Beliefs*, 442.

40. Mary Wright interview, *SN*, vol. 7, *Kentucky Narratives*, 64; "Lidia Jones," in *The American Slave: A Composite Autobiography*, ed. George P. Rawick, supp. ser. 2, vol. 1, *Alabama, Arizona, Arkansas, District of Columbia, Florida, Georgia, Indiana, Kansas, Maryland, Nebraska, New York, North Carolina, Oklahoma, Rhode Island, South Carolina, Washington Narratives* (Westport, CT: Greenwood, 1979), 98.

41. Perdue, Barden, and Phillips, *Weevils in the Wheat*, 249; Puckett, *Folk Beliefs*, 344.

42. "Mrs. Rush," *Am. Sl.*, supp. ser. 1, vol. 4, pt. 2, *Georgia Narratives*, 649–52; "Robert Heard," ibid., 170–71; "Aunt Clussey," *Am. Sl.*, supp. ser. 1, vol. 1, *Alabama Narratives*, 20; "Casie Jones Brown," *Am. Sl.*, vol. 2, pt. 1, *Arkansas Narratives*, 271; "Hamp Kennedy," *Am. Sl.*, vol. 9, *Mississippi Narratives*, 84–90. Other interviewees who talked about covering mirrors at the time of death were John Cole (*SN*, vol. 4, *Georgia Narratives, Part 1*, 230); Robert Henry (*SN*, vol. 4, *Georgia Narratives, Part 2*, 198); "Elsie Moreland" (*Am. Sl.*, supp. ser. 1, vol. 4, pt. 2, *Georgia Narratives*, 453); "Caroline Malloy" (*Am. Sl.*, supp. ser. 1., vol. 4, pt. 2, *Georgia Narratives*, 415); "Easter Reed" (*Am. Sl.*, supp. ser. 1, vol. 4, pt. 2, *Georgia Narratives*, 502); and "Rena Clark" (*Am. Sl.*, supp. ser. 1, vol. 7, pt. 2, *Mississippi Narratives*, 408–22).

43. Puckett, *Folk Beliefs*, 81–82. The mirror described as having two defects that resembled human eyes could be easily explained by how accurately reflective looking glasses aged in circular patches of gray. See chapter 1n17 herein.

44. Georgia Writers' Project, *Drums and Shadows*, 20–21.

45. "Mary Jane Simmons," *Am. Sl.*, supp. ser. 1, vol. 4, pt. 2, *Georgia Narratives*, 561–70. These questions appeared with the answers given by Mary Jane Simmons, whose interview is filed alphabetically with the other 312 Georgia interviews. See John D. Fair, "The Georgia Slave Narratives: A Historical Conundrum," *Journal of the Historical Society* (2010): 252.

46. "Elsie Moreland," *Am. Sl.*, supp. ser. 1, vol. 4, pt. 2, *Georgia Narratives*, 458; "Mary Jane Simmons," *Am. Sl.*, supp. ser. 1, vol. 4, pt. 2, *Georgia Narratives*, 562; Robert Henry interview, *SN*, vol. 4, *Georgia Narratives, Part 2*, 197–98.

47. Perdue, Barden, and Phillips, *Weevils in the Wheat*, 155.

48. See Raymond J. DeMallie, ed., *Handbook of North American Indians: Plains*, vol. 13, pt. 1 (Washington: Smithsonian Institution Press, 2001), 10.

49. Norman Feder, *Art of the Eastern Plains Indians: The Nathan Sturges Jarvis Collection* (Brooklyn, NY: Brooklyn Museum, 1964), 32; Frances Eyman, "An Unusual Winnebago War Club and an American Water Monster," *Expedition* 5, no. 4 (1963): 31–35.

50. Grace Lee Nute, "Peter Rindisbacher, Artist," *Minnesota History Magazine* 14, no. 3 (1933): 283–87.

51. This might also explain, at least in part, the popularity among Native Americans of silver pieces of jewelry and body art. See Martha Hamilton, *Silver in the Fur Trade, 1680–1820* (Chelmsford, MA: Martha Hamilton Publishing, 1995).

52. Joseph Epes Brown, "The Unlikely Associates, a Study in Oglala Sioux Magic and Metaphysic," *Ethnos* 35 (1970): 13.

53. Brown, "Unlikely Associates," 12–13; DeMallie, *Handbook of North American Indians: Plains*, vol. 13, pt. 2, 808.

54. Alice C. Fletcher, "The Elk Mystery or Festival of the Ogllala Sioux," in *Annual Report of the Peabody Museum of American Archaeology and Ethnology*, vol. 3, *1884–1886* (Salem, MA: Salem Press, 1887), 276–88.

55. Ibid., 286.

56. Amos Bad Heart Bull (Drawings) and Helen H. Blish (Text), *A Pictographic History of the Oglala Sioux* (Lincoln: University of Nebraska Press, 1967), 200–201.

57. Ibid.

58. S. Elizabeth Bird has succinctly criticized the work of these "salvage ethnographers" as profoundly influenced by their beliefs about Native cultures. As a result, they "produced accounts of peoples programmed by cultural rules, calmly going about their (ultimately doomed) business." Westerners have often described other societies they deemed primitive as overly controlled by cultural rules (as I discussed in my introduction). S. Elizabeth Bird, "Savage Desires: The Gendered Construction of the American Indian in Popular Media," in *Selling the Indian: Commercializing and Appropriating American Indian Cultures*, ed. Carter Jones Meyer and Diana Royer (Tucson: University of Arizona Press, 2001), 63.

59. Clark Wissler, "The Whirlwind and the Elk in the Mythology of the Dakota," *Journal of American Folklore* 18, no. 71 (1905): 257, 266–67.

60. This 1830s Eastern Sioux wooden fan can be seen at "The Plains Indians: Artists of Earth and Sky," Metropolitan Museum of Art, http://metmuseum.org/exhibitions/view?exhibitionId=%7B254A181E-CA25-4BC9-B15A-A167688D711B%7D&oid=641568. Bad Heart Bull and Blish, *Pictographic History*, 201; Clark Wissler, "Some Protective Designs of the Dakota," in *Anthropological Papers of the American Museum of Natural History*, vol. 1, pt. 2 (New York: Order of the Trustees, 1907), 40.

61. See James H. Howard, "Notes on the Dakota Grass Dance," *Southwestern Journal of Anthropology* 7, no. 1 (1951): 84.

62. DeMallie, *Handbook of North American Indians: Plains*, vol. 13, pt. 2, 807. This account of the Sun Dance is from Frances Densmore, *Teton Sioux Music* (Washington: Government Printing Office, 1918), 149. According to Clyde Holler, Densmore's account is "authoritative, professional." See Clyde Holler, *Black Elk's Religion: The Sun Dance and Lakota Catholicism* (Syracuse, NY: Syracuse University Press, 1995), 101.

63. "Expedition Notes," American Journeys, www.americanjourneys.org/aj-144b; Edwin James, *An Account of an Expedition from Pittsburgh to the Rocky Mountains, Performed in the Years 1819–1820*, vol. 2 (London: Longman, Hurst, Rees, Orme, and Brown, 1823), 114, 248–49, 300.

64. A ca. 1880 horse mirror board from the Ioway can be seen at *A Song for the Horse Nation: Horses in Native American Cultures*, Smithsonian Museum of the American Indian, http://nmai.si.edu/exhibitions/horsenation/carvings.html. See also David W. Penney and Janet Stouffer, "Horse Imagery in Native American Art," *Bulletin of the Detroit Institute of Arts* 62, no. 1 (1986): 18–25.

65. DeMallie, *Handbook of North American Indians: Plains*, vol. 13, pt. 2, 759.

66. Densmore, *Teton Sioux Music*, 248–49; William Wildschut, "Arapaho Medicine-Mirror," *Indian Notes* 4, no. 3 (1972): 252–57.

67. Clark Wissler, "Decorative Art of the Sioux Indians," *Bulletin of the American Museum of Natural History*, vol. 18, pt. 3 (1904): 231–78, 238.

68. Philip Deloria's *Indians in Unexpected Places* considers how Native peoples embraced and engaged with modernity in the late nineteenth- and early twentieth-century United States. Deloria argues that "as Native peoples have lived lives that refused the white expectation that they would have an inferior engagement with technology, they have also sought to represent such lives, portraying the ways in which Indian people have created distinctly Native spaces that are themselves modern." Although in the earlier period under study here Indians had fewer opportunities to shape their own image for a larger white audience, they certainly lived their lives in ways that engaged the mirror on their own terms and demonstrated, as Deloria concludes was true of the later period as well, "that the entire world of the modern belonged—and belongs—to Indian people, as much as it does to anyone else." Philip J. Deloria, *Indians in Unexpected Places* (Lawrence: University Press of Kansas, 2004), 156, 232.

Epilogue

1. Mary Ames, *From a New England Woman's Diary in Dixie in 1865* (Norwood, MA: Plimpton Press, 1906), 2, 70–72.

2. Whites were disgusted with and threatened by how African Americans appropriated many items of material culture—not just mirrors—following the Civil War in order to express their new status. In 1865, for example, Sidney Andrews, a northern journalist who traveled in North Carolina, quoted a white woman to have said that "the chief ambition of a wench seems to be to wear a veil and carry a parasol." Sidney Andrews, *The South since the War: As Shown by Fourteen Weeks of Travel and Observation in Georgia and the Carolinas* (Boston: Ticknor and Fields, 1866), 186, https://archive.org/stream/southsincewarassooandr#page/n5/mode/2up.

3. "Humors of the Day," *Harper's Weekly*, Sept. 1, 1866, 551.

4. Not everyone who practiced the Ghost Dance believed it should be connected with violent conflict, including Ghost Dance originator Paiute Jack Wilson (Wovoka). See Alice Beck Kehoe, *The Ghost Dance: Ethnohistory and Revitalization*, 2nd ed. (Long Grove, IL: Waveland, 2006), 13–14; Gregory E. Smoak, *Ghost Dances and Identity: Prophetic Religion and American Indian Ethnogenesis in the Nineteenth Century* (Berkeley: University of California Press, 2006), 1–2; and Raymond J. DeMallie, "The Lakota Ghost Dance: An Ethnohistorical Account," *Pacific Historical Review* 51, no. 4 (1982): 385–405, 387.

5. See Alfred L. Kroeber, "The Arapaho," *Bulletin of the American Museum of Natural History* 18, pt. 4 (1907): 356–57, https://books.google.com/books?id=E-MyAQAAMAAJ&. See figure 127 (356) for a drawing of one of these mirrors.

Index

Page numbers in *italics* indicate figures.